SONY FS7

QUICK START AND BASIC REFERENCE

(based on Firmware Version 4.20)

SONY FS7

QUICK START AND BASIC REFERENCE
(based on Firmware Version 4.20)

KAHALA PRESS LLC   HONOLULU

Copyright © 2018 by Greg Keast

All rights reserved except any material held in copyright by others. Any such material is acknowledged and for educational purposes only. The opinions expressed in this book are solely those of the author and without warranty. SONY is a registered trademark of the Sony Corporation. No such use, or the use of any trade name, is intended to convey endorsement or affiliation with this book.

ISBN-13: 978-1986917407
ISBN-10: 1986917401

PREFACE

The goal of this book is to not only get you started with using the FS7 quickly but also to serve as a basic reference that can be used in the field. It is not meant to answer every question you might have, review every feature, or analyze every combination of settings. This book is for quick start and basic reference only. It is designed to supplement the Sony manual, not replace it. The book and the manual share information, yet there are major differences between them not only in terms of content but in how that content is organized and presented.

Many features of the camera are self-explanatory but given the menu system and the camera's layout, it can be a challenge to know where to start, and it can take time to develop a solid workflow. This book, through a series of checklists, is meant to address those issues. However, in using the checklists, please note that if a feature or setting is not mentioned, then that means the default setting is considered the best setting for most situations. If you are interested in a setting not included, then you should consult the manual or perform an Internet search, which will usually lead you to one of the main forums devoted to the camera. Those resources, as well as others, are listed in the back of the book.

This book structures itself around three pre-shooting checklists.

The first is the Quick Start/Initial checklist, and it is designed for new users; however, everyone should go through this checklist at least once, especially in the beginning. The list has one-time-only settings that are not adjusted again in any other checklist.

There are two other major checklists and which you use depends upon the base shooting mode you choose. The FS7 has two.

The first is Custom mode and is meant for shooters who want to get the footage right in the camera and don't plan to do much, if any, color correction later. This is a perfectly valid choice that can yield professional results.

The second mode is CINE-EI and *requires* color correction after shooting is done. It is for those filmmakers who plan to push the footage as far as they can creatively in post-production. However, if you are not adept at color correction and plan to shoot in CINE-EI, then you will need to learn basic color correction or plan to have someone else do that for you. Note: As a bonus, there is a short jump-start section on color correction at the end of the book.

The checklists are designed so if you don't understand a particular setting or want to learn more about it, you can reference its assigned page number. Also, please keep in mind that the checklists represent the preferences of the author and might not be the same as yours. For this reason, if you see a step on a checklist and you don't feel it is necessary or important, then by all means skip it. If you want to add a step or modify one, then that's perfectly fine too.

The checklists are only guidelines and do not mean there is only one way of doing things or getting the camera ready. There are many roads to Rome, and any recommendations are simply that—recommendations. I encourage you to keep studying and researching things for yourself.

It is expected that once you gain knowledge and experience using the camera, you will create your own custom checklists (or modify ours) and edit the User menu to best meet your needs. However, once you edit the User menu, then the checklists in this book will appear out of order and will be harder to use.

My primary goal is to make using the FS7 an easier and more enjoyable process. Great care was taken to ensure the accuracy of the information in this guide; however, if you happen to notice any errors or have any suggestions for improvement, please contact us at:

admin@diydigitalcinema.com.

Greg Keast
May 2018

CONTENTS

PREFACE	iii	FIRMWARE	48
INTRODUCTION	7	FLICKER	49
QUICK-START / INITIAL LIST	10	FOCUS	49
CUSTOM-MODE CHECKLIST	12	FOCUS ASSIST	50
CINE-EI CHECKLIST	14	FORMAT MEDIA	50
CHANGING CHECKLIST	15	GAMMA	51
ALL FILE	16	HIGH / LOW KEY	55
APERTURE	17	HOLD FUNCTION	55
APR	17	INTERVAL (TIME-LAPSE)	56
ASSIGNABLE BUTTON	18	ISO / GAIN / EI	57
AUDIO	19	ISO / GAIN / EI SWITCH	63
AUDIO LEVELS	22	LANGUAGE	64
AUDIO OUTPUT	23	LENS / CAP / HOOD	64
AUTO BLACK BALANCE	24	LOOK PROFILE	65
AUTO FOCUS	24	MAPPING LUTS	67
AUTO SETTINGS	25	MATRIX	69
BASE SHOOTING MODE	26	MONITOR LUT (MLUT)	70
BATTERIES	27	MULTI-MATRIX	71
BLACK LEVEL	28	NOISE SUPPRESSION	73
CALIBRATION	29	PAINT	74
CAMERA OUTPUTS	30	PEAKING	77
CHANNEL 1 & 2 CONTROL	30	POWER	78
CINE-EI MODE	31	PRESET WHITE BALANCE	79
CLOCK SET	32	REC FORMAT	80
CODEC	32	RECORD	80
COLOR SPACE	33	S & Q	81
COUNTRY (NTSC/PAL)	34	S-GAMUT/SLOG2	33
CUSTOM MODE	35	S-GAMUT3/SLOG3	33
CUSTOM WHITE BALANCE	36	S-GAMUT3.CINE/SLOG3	33
DISPLAY	39	SCAN MODE	83
EXPOSURE	41	SCENE FILES	84

SHUTTER	85	XQD MEMORY CARDS	92
SLOT SELECTION	86	ZEBRA FUNCTION	93
SWITCHING MODES	87	BASIC CORRECTION	94
SYSTEM RESET	87	ONLINE RESOURCES	107
VF (VIEWFINDER)	88	DEFINITIONS	108
VIDEO	89	CHECKLISTS	112
VIDEO SIGNAL MONITOR	91	INDEX	118
WHITE CLIP	91	NOTES	121

Sony marketing brochure for the FS7 circa 2015.

INTRODUCTION

The pre-shooting checklists are designed to follow the order in which the settings appear in the FS7's default menu system. The only exceptions to this are some first-time settings such as the hold switch, language, and clock. In these instances, you should jump ahead to adjust those settings, then return to the checklist to follow the menu's structure in a sequential, one-step-at-a-time fashion. In the camera's default setup, there are some settings that will appear more than once in the menu. Sometimes the setting will be adjusted the first time it comes up, and sometimes it will be set later. For consistency, it is preferred that some settings be adjusted later in their regularly assigned menu group versus in the User menu. The User menu is an easily changed assortment of settings, but by default is already set up for you—courtesy of Sony.

The Quick Start / Initial Checklist is designed a little differently from the other checklists. It assumes you are new to the camera and don't know where everything is, so it provides more guidance. It is expected that it will take you awhile to make your way through the setup the first time, especially if you need to read up on settings to learn more about them. However, once you are familiar with the camera and know where things are, then it is expected that you will graduate to the other checklists. The other checklists assume that you know where things are and those checklists are meant to serve more as reminders.

The menu system has twelve primary sections.

SONY MENU SYSTEM
LEVEL-ONE

USER	User-defined menu items; by default, Sony has populated this menu with some of the most common camera settings.
CAMERA	Major camera settings such as ISO, shutter speed, and noise suppression.
PAINT	A multitude of color-control settings to bake-in a custom image.

AUDIO	Audio input and output settings and controls.
VIDEO	Video input and output settings.
VF	Viewfinder settings for exposure tools such as zebras and focus peaking. Some of these settings can also be controlled on the viewfinder itself.
TC / UB	Timecode and user-bit settings.
RECORDING	Slow motion, time-lapse, video cache, and other recording options.
THUMBNAIL	Video-clip management and display.
MEDIA	Media and file-naming settings.
FILE	File management including scene files and user-defined gamma settings.
SYSTEM	Controls many basic camera operations and features.

Inside each section, there are two levels of additional menus that can be accessed.

Pressing the SEL/SET wheel allows you to select an item and pressing the Back button allows you to escape one level back until you reach the first-level menu. Pressing the MENU button allows you to enter the first level of the menu and pressing the MENU button again allows you to exit the menu completely. With practice, you will learn to navigate the system quickly.

Many settings and features on the camera can only be accessed internally through the Menu system. However, some internal Menu settings can also be accessed externally while a few settings can only be accessed externally. For instance, the Auto or Manual setting for audio channels 1 and 2 can only be accessed externally. Most of the time, you will be changing settings through the internal menu.

It is strongly recommended that you keep using the checklists until you feel comfortable navigating the menu and know where everything is located. And even after you have used the camera for a while, it is still recommended that you use some type of checklist, whether it is yours or ours, to make sure you don't miss something.

Since there is no way of knowing which accessories you might be equipping the camera with, the checklists assume a setup using the camera as it comes packaged with no additional accessories such as monitors or microphones.

If you edit the User menu, which you might ultimately decide to do, then be aware that the checklists will be harder to follow because you will have changed the default User-menu system on which the checklists are based. For this reason, you should not edit the User menu system until you are thoroughly familiar with the camera and have your own system developed. If you do edit the User menu, then you can change it back to its default layout by resetting the camera. The reset function can be accessed through the Systems section of the menu.

QUICK TIP
No matter how far you are into the menu system, pressing the CANCEL/BACK button THREE TIMES and the MENU button ONCE will exit you from the menu.

QUICK START / INITIAL CHECKLIST

1. Install lens and leave the cap on. [64]
2. Install XQD card(s). [92]
3. Install fully charged battery. [27]
4. Power on. [78]
5. Undo the Hold setting if on. [55]
6. If APR activates, stop down the iris and execute it. [17]
7. Jump to: System > Version---Is firmware current? 4.20 [48]
8. System > All Reset---Reset camera. [87]
9. System > HOLD Switch Setting > w/ Rec---Turn off. [55]
10. System > Language: English---Change if needed. [64]
11. System > Clock Set---Change as needed. [32]
12. Turn off all Auto Settings. [25]
13. Turn off Auto Focus. [24]
14. Return to: User > Country---Set PAL or NTSC. [34]
15. User > Base Setting > Shooting Mode---Set to Custom. [26]
16. User > Rec Format---Select format. [80]
17. User > Codec---Select. [32]
18. User > Assignable Buttons---Assign buttons 2 and 3. [18]
19. User > Format Media---Format. [50]
20. External: Confirm Slot Selection. [86]
21. User > Gamma---Select.* Check chart, note IRE level. [51]
22. User > Aperture---Turn on. Set to +20. [17]
23. User > Peaking---(Optional setting.) [77]
24. User > ISO / Gain---Set ISO or dB. Set values L, M, & H. [57]
25. User > Zebra---Set Zebra 1 based on gamma. [93]
26. Camera > Shutter---Set mode. Set speed/angle. [85]
27. Camera > Color Bars > Setting ---Turn on, then choose: Type > SMPTE---Calibrate contrast and brightness. [29]
28. Camera > Auto Black Balance---Execute. [24]
29. Paint > Black Level---Set to -3. [28]
30. Paint > White Clip---Set to 109. (Disabled if S-Log is on.) [91]
31. External: AUTO/MAN (Audio)---Set to Manual (CH 1 & 2). [30]
32. Audio > Audio Inputs---Set inputs, wind filter on, side level. [19]

33. Audio > Audio Output > Monitor Volume---Adjust. [23]
34. VF > Video Signal Monitor > Setting---Choose waveform. [88]
35. Check display on VF to confirm all settings. [39]
36. Remove lens cap. [64]
37. Make sure lens is clean, free from dust and smudges. [64]
38. Install lens hood or matte box. [64]
39. Confirm the ISO switch is set to L. [63]
40. Adjust exposure. [41]
41. Set WB switch to A or B. Perform custom WB. [36]
42. Adjust focus. [49]
43. Check audio levels on Display or Status page. [22]
44. Put settings on hold to prevent changes. (optional) [55]
45. Ready to record. [80]

*If you reset the camera, as is recommended, the gamma curve should be set to STD 5, which is a good setting to try when testing out the camera for the first time. After you have gone through the setup at least once, then you can experiment with other gamma curves.

BE SURE THE POWER IS OFF BEFORE STORING THE CAMERA.

CUSTOM-MODE CHECKLIST

1. Install lens and leave the cap on. [64]
2. Install XQD card(s). [92]
3. Install fully charged battery. [27]
4. Power on. [78]
5. Undo the Hold setting if on. [55]
6. If APR activates, stop down the iris and execute it. [17]
7. Turn off all Auto Settings. [25]
8. Turn off Auto Focus. [24]
9. Set: Base Setting > Shooting Mode > Custom. [26]
10. Set: Base Setting > Scan Mode. Change if needed. [83]
11. Set: Rec Format, take mental note. [80]
12. Set: Codec. [32]
13. Format Media. [50]
14. Confirm Slot Selection. [86]
15. *If loading a scene file*:
    Jump to: File > Scene File > Scene White Data > Turn off. [84]
    Then load the scene file and return to User menu. [84]
16. **Gamma*** : Be sure to remember what this setting is. [41,52]
17. Aperture > On > Set to 10-20. [17]
18. Peaking...Optional. [77]
19. **ISO / Gain / EI *** : Set values for L, M, & H. [57]
20. Set Zebra 1 based on gamma. [93]
21. **Shutter speed / angle. *** [85]
22. **Noise suppression.*** Set to high in low light. [73]
23. Auto Black Balance. Execute. [24]
24. **Black*** : -3 (optional) [28]
25. **White Clip*** : 109 [91]
26. **Paint > Matrix > Setting* :** On [69]
27. **Paint > Matrix > Adaptive Matrix*** : On [69]
28. **Paint > Matrix Preset*** : (optional) [69]
29. **Paint > Preset Select*** : (optional) [74]
30. **Paint > User Matrix*** : (optional) [75]
31. **Paint > Matrix User > Level*** : (optional) [75]

32. **Paint > Multi-Matrix > Setting *** : (optional) [71]
33. **Paint > Multi-Matrix*** : YL & R set to -20 to -30. (optional) [71]
34. Audio Switch: Set to Manual. [30]
35. Audio: Set inputs, wind filters on, side level. [19]
36. Audio Output> Adjust monitor volume. [23]
37. VF > Video Signal Monitor > Setting > Waveform. [88]
38. Use display to confirm all settings. [39]
39. Remove lens cap. [64]
40. Make sure lens is clean, free from dust and smudges. [64]
41. Install lens hood. [64]
42. Confirm the ISO switch is set to L. [63]
43. Adjust exposure. [43]
44. Set WB switch to A or B. Perform custom WB. [36]
45. Adjust focus. [49]
46. Check audio levels. Display or status. [22]
47. Put settings on hold to prevent changes. (optional) [55]
48. Ready to record. [80]

* *If a setting on the checklist appears in **BOLD** and is marked by an asterisk, then this means the setting will change to whatever value is stored on a loaded scene file. If you already know what the scene file settings are, then you can skip all the settings in **bold** because they will be set by the scene file.*

If you are not certain what all the scene file settings are, then do not skip these settings. You will need to confirm what they are. For instance, a scene file will overwrite your shutter speed. This is usually a setting you will not want changed by the scene file unless it is a scene file you created for yourself.

If you find a scene-file value you wish to change, you can change it. The setting you enter will stay in effect until you load another scene file.

BE SURE THE POWER IS OFF BEFORE STORING THE CAMERA.

You make a film for yourself. That is an absolute necessity.
                                              JEAN-PIERRE JEUNET

CINE-EI CHECKLIST

1. Install lens and leave the cap on. [64]
2. Install XQD card(s). [92]
3. Install fully charged battery. [27]
4. Power on. [78]
5. Undo the hold setting if on. [55]
6. If APR activates, stop down and execute. [17]
7. Turn off all Auto Settings. [25]
8. Turn off Auto Focus. [24]
9. Set: Base Setting > Shooting Mode > CINE-EI. [26]
10. Set: Base Setting > Color Space. Choose one. [33]
11. Set: Base Setting > Scan Mode. Choose Center to zoom. [83]
12. Set: Rec Format, take mental note. [80]
13. Set: Codec [32]
14. Format Media [50]
15. Confirm Slot Selection. [86]
16. ISO / Gain / EI : Set values for L, M, & H. [57]
17. Set zebra based on gamma [93]
18. Set: Shutter speed or angle. [85]
19. Noise suppression. Set to high if low light. [73]
20. Execute: Auto Black Balance [24]
21. Audio Switch: Set to Manual [30]
22. Audio: Internal, Wind, Side Level [19]
23. Audio Output> Monitor Volume, Headphone Out, Output [23]
24. Video > Monitor LUT > SDI1 & Internal > Off. Confirm off. [89]
25. Video > Viewfinder > MLUT > Turn on. [89]
26. Set: Video Signal Monitor > Setting > Waveform. [91]
27. Set: Video Signal Monitor > Source > SDI2. [91]
28. Use display to confirm all settings. [39]
29. Remove lens cap. [64]
30. Make sure lens is clean, free from dust and smudges. [64]
31. Install lens hood. [64]
32. Confirm the ISO switch is set to L. [63]

33. Adjust exposure [44]
34. Set white balance closest to actual lighting. [79]
35. Adjust focus. [49]
36. Check audio levels. Display or status. [22]
37. Put settings on hold to prevent changes. (optional) [55]
38. Ready to record. [80]

BE SURE THE POWER IS OFF BEFORE STORING THE CAMERA.

CHANGED LIGHTING CHECKLIST*

1. Double-check shutter speed, frame rate, WB, and ISO. [39]
2. Confirm battery power and media card space. [27, 39]
3. Check lens. [64]
4. Set exposure based on gamma used. [41]
5. White balance [36,69]
6. Check focus. [49]
7. Check audio levels. [22]
8. Ready to record. [80]

This checklist is when you have changed locations or the lighting has changed, and there is no need to change the rest of the settings.

NOTE:
When you are using the FS7's menu system, if you encounter a setting that is dimmed or receive the message: CANNOT PROCEED, then that means a feature is disabled due to the base mode you are shooting in or another setting overrides it. This is a normal message to get and serves as a reminder.

ALL FILE

The ALL FILE function allows you to save nearly every setting on the camera in a file that can be loaded into the camera. It is a useful feature because once you have your camera setup just the way you like it, you can save the settings and know that you can restore the same setup by loading a single file. The function can store 64 camera profiles, which is more than you will ever need.

PROCEDURE

1. If you don't already have an SD card in the camera, then insert one and format it.

2. FILE > ALL FILE > FILE ID
Navigate to FILE ID and give the file a name. Your camera's settings cannot be saved in an ALL FILE until it is given a name. Select DONE when finished.

3. FILE > ALL FILE > SAVE SD CARD
Choose a blank location and Execute the save function.

To load a camera profile, simply navigate to:
FILE > ALL FILE > LOAD SD CARD
….then choose the file you want to load.

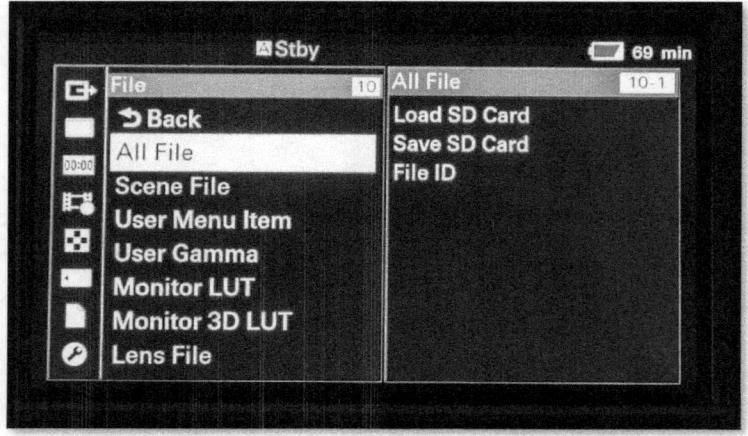

APERTURE (Optional / Custom Mode only)

Aperture applies a sharpening effect (a high-end frequency boost) that can enhance textures, fine lines, and details.

If you find that your images appear soft, you can enable this setting.

The consensus among FS7 users is that between the detail and aperture settings, aperture does a better job of sharpening; therefore, the recommendation is to leave the detail function off and set aperture between 10 and 20. Too much sharpening can lead to unwanted artifacts, so going beyond 20 is not suggested.

Some FS7 users report they can get sharp imagery without any aperture or detail applied while others always use it. There is no right or wrong here, so the best advice is to shoot some test footage to see for yourself if you notice any appreciable difference.

It should also be noted that sharpening can also be applied in post if necessary.

Aperture is one of those settings that will appear again in the Menu system, so when you see it again under the Paint section, you can ignore it as it will have already been set.

APR (Automatic Pixel Restoration)

This is a function that can automatically appear during system start-up. If it does, it is suggested that you execute the function but when you do, be sure to leave the lens cap on and stop down the iris.

APR is an automatic process that is supposed to assess the sensor for any stuck or out-of-spec pixels. A dead pixel will be black. A stuck pixel can be white, green, blue, or red. If it finds any, it is supposed to map them out with values from adjacent pixels. The result is an image with no dead or distracting pixels.

ASSIGNABLE BUTTON

There are three assignable buttons on the camera body. If the arm grip is attached, then you have three more.

USER BUTTONS ASSIGNED BY DEFAULT (CAMERA BODY)	
BUTTON 1	Activates S & Q function (slow motion).
BUTTON 2	Activates Auto Iris function.
BUTTON 3	Displays the User menu on the viewfinder.

If you go to **System > Assignable Button**, you can reassign functions to each of the buttons.

For instance, if double-checking your focus is important to you, you can assign Button 3 to Focus Magnifier x4/x8. This allows you to zoom in on your subject to ensure your focus is as sharp as it can be. But then again, this is all about preferences, so you could assign it to any of the many functions available. You should also consider assigning High Key / Low Key Function to Button 2, the IRIS function.

AUDIO

For a quick-start or first-time setup, it is assumed no additional microphones are being attached. If you choose to add external microphones, then those menu options are in the same location as the internal microphone.

For a first-time setup using the camera's built-in functions, you can enable the internal microphone just to check out its functionality. If you were to attach an external microphone with an XLR cable, you would set it to INPUT 1 and plug the microphone cable in Input 1 on the right side of the camera.

You can also adjust your external microphone or input to LINE, MIC, and MIC+48V (Phantom power).

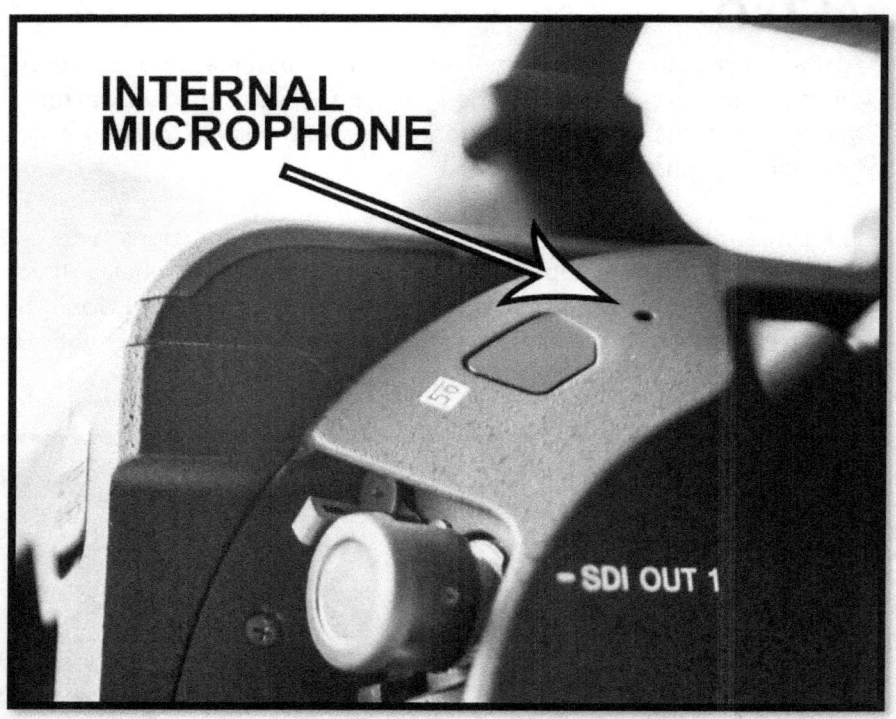

The internal microphone is tiny and at the back of the camera. It is not built to record professional audio and at best, it is for reference or testing purposes only.

```
AUDIO INPUT > CH 1 INPUT SEL. > INTERNAL MIC
AUDIO INPUT > CH 2 INPUT SEL. > INTERNAL MIC
```

To enable the internal microphone for the initial setup, you enable it from the Audio Input submenu.

```
AUDIO INPUT > CH 1  WIND FILTER >  ON
AUDIO INPUT > CH 2  WIND FILTER >  ON
```

Further, if using the internal microphone or an external microphone, then turning on the wind filters is recommended. By default, the filters are off. They are set from the Audio Input menu as well.

AUDIO INPUT > CH1 LEVEL > SIDE
AUDIO INPUT > CH2 LEVEL > SIDE

And finally, under CH1 and CH2 level, it is recommended that you change this to SIDE. This way if you want to boost the gain of the audio-input signal coming from the microphone, you can easily adjust the level from the dials on the left side of the camera. By default, these dials should be set to their default positions, which is 5 on the dials. If you have a hard time adjusting the dials, they can be fully accessed by pulling outward on the side door of the camera. However, it should rarely be necessary to increase the audio gain that much and doing so is not usually recommended. If you are using an external microphone and need to boost the signal, you are encouraged to use a quality preamplifier like the Sound Devices MP-1 or move the microphone closer to the source of the sound.

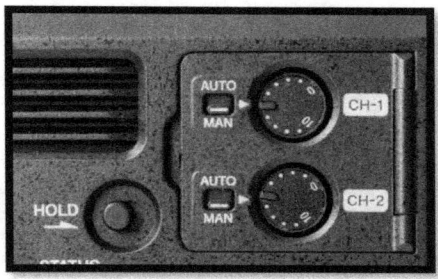

Please note that if the external audio switches are set to Manual, then automatic gain control or AGC will be disabled, so adjusting that setting on the menu makes no difference. It is not recommended that you use automatic gain control under any circumstance.

You will probably not want to record anything seriously important using the internal microphone except reference or scratch-track audio. If you are capturing audio into the camera for a project, you would have to use an external microphone. With the use of an external microphone, you can usually capture good-quality audio from the FS7 without a preamplifier. On lesser cameras, a preamplifier is usually required, but the built-in recorder on the FS7 has a good signal-to-noise ratio and low-noise floor.

AUDIO LEVELS

Right before you are ready to start recording, you should double-check the audio levels one last time. This can be done easily by pressing the Status button and turning the Selection wheel to the Audio Status page.

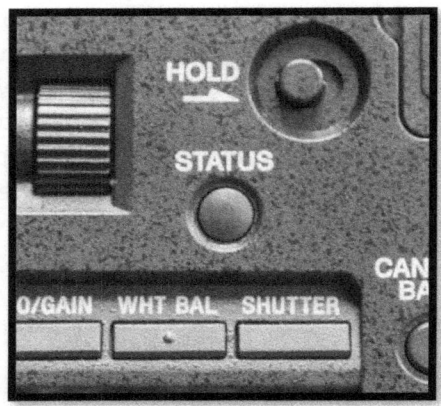

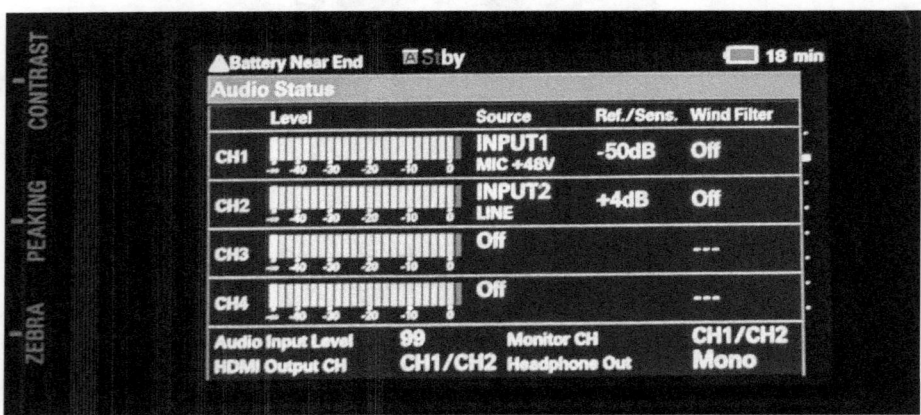

The FS7 records high-resolution audio at 24-bit/48kHz.

Audio levels hovering between -20 and -10 are sufficient. Assuming that good recording techniques are being used, you should be able to boost the audio in post with no appreciable increase in the noise level

AUDIO OUTPUT

AUDIO OUTPUT > MONITOR VOLUME > 0-15

If you want to adjust the volume on your audio headphones, then you need to go all the way into this section of the menu to do that. The headphone jack is at the back of the camera on the upper left side.

The headphone input accepts a standard 3.5 mm jack.

> *Sound is half the picture.*
> **GEORGE LUCAS**

AUTO BLACK BALANCE

This is a function you should execute regularly. Situations that call for black balancing include:

* Shooting in changing temperatures.
* Using the camera for the first time.
* The camera has been in storage for some time.
* You notice excessive grain or noise.

It is recommended that you do the black balance after the camera has warmed up and do it in a dark location with the iris stopped down and the lens cap on. Black balance helps adjust the black levels of all the pixels on the sensor.

AUTO FOCUS

It is advisable to have your focus control on manual. You don't want the camera deciding what to focus on. You, as the operator, will want to control that. On the FS7, under FOCUS on the front left side of the camera, you can switch from Auto to Manual focus.

If you want to use Auto Focus to quickly focus on something, then you can press the Focus/Push Auto button, and it will initiate auto focus. Once released, the camera will revert to manual mode.

AUTO SETTINGS

By default, the auto settings should be off. However, by looking at the display screen, you can see if any of them are on.

If you press the FULL AUTO button on the left side of the camera, you will see several Auto Exposure settings appear as activated on the display screen. If you press the button again, it will clear and deactivate them.

As a rule, you don't want any automatic settings on. You want to be in complete control over the camera. The automatic settings include the following:

PRIMARY AUTO SETTINGS		
Automatic Shutter	Auto Gain Control	Auto Iris
Auto Exposure	Auto White Balance	

Note: Auto settings can only be turned off and on in Custom Mode. If you are shooting in CINE-EI, auto exposure settings are disabled.

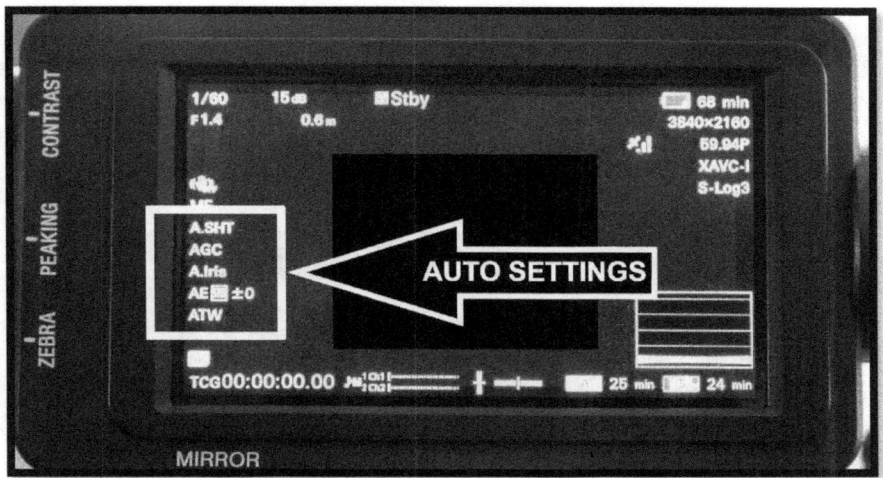

BASE SHOOTING MODE

This is the most important setting on the menu. Shooting mode controls whether the camera will be operating in CINE-EI or Custom mode. Both modes profoundly affect the camera and will disable or enable a wide range of other controls and settings. Which mode you choose will be determined by the nature of your project and how much color correction you are willing to do in post. If you want to shoot footage that will be *good-to-go* right out of the camera, then you should choose Custom mode. If you want to shoot an image that you will have maximum creative control of in post and are okay with color correction, then shoot in CINE-EI.

> USER > BASE SETTING > SHOOTING MODE > CUSTOM
> CINE-EI

If you choose Custom mode, then you will notice that under the next option, Color Space, you have no choice of color space, and it will default to Matrix mode for color space; however, if you choose CINE-EI, then you will see that you have a choice of three color spaces. If you try to make a change on a setting that is disabled because of the mode you are shooting in, you will receive the message: "Cannot proceed."

If you properly expose footage in CINE-EI and are proficient with grading, you will eventually discover that you can make CINE-EI footage look like anything you can shoot in Custom mode. The same cannot be said of Custom mode. If you shoot in Custom mode, the look is baked-in and while color correction can be done, the footage can start to break down.

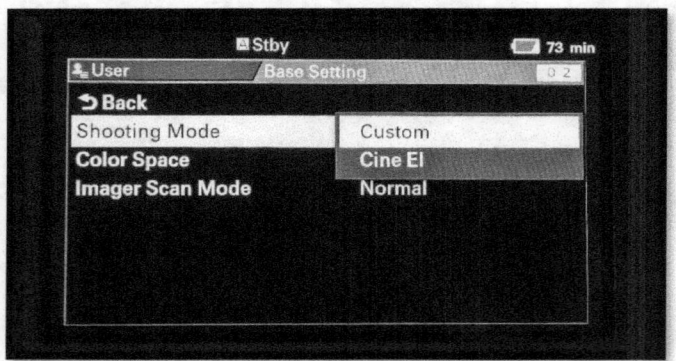

BATTERIES

For any camera under warranty, using aftermarket batteries is not recommended. However, there are many FS7 users who use aftermarket products and report good results. The decision to go with aftermarket products is solely yours to make.

While many users install the optional extension unit to use V-mount batteries, there is nothing wrong with the proprietary batteries that Sony sells for the camera. These are known as the BP series and there are three:

SONY BATTERY LIFE	
TYPE	MINUTES
BP-U30	~ 60
BP-U60 (T)*	~ 120
BP-U90	~ 180

* T version of this battery includes Hirose Power Tap
~ approximate values

These units are fully compatible with the camera, which means information about the battery and its status is monitored and displayed via the Status screen.

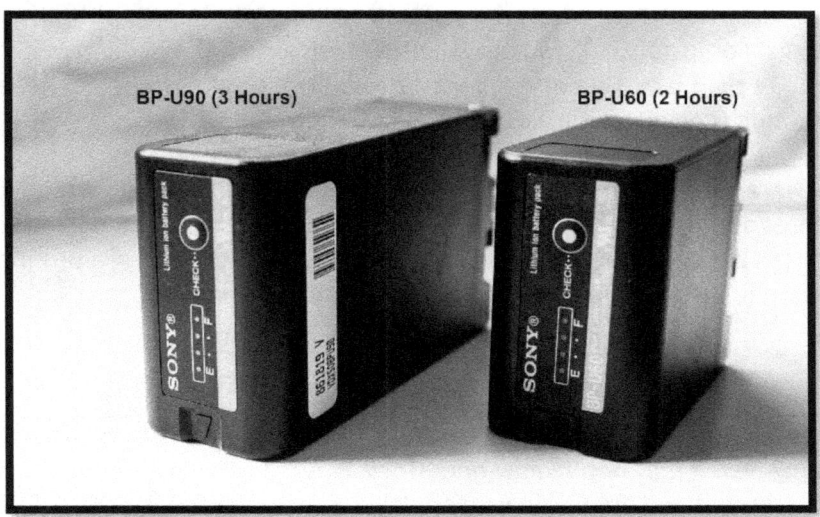

If you want to use V-Mount type batteries, then you would have to purchase the Sony XDCA-FS7 but as of 2018 that alone retails for nearly $2,000. There are less expensive mounts available and those start around $400 and up, not including the batteries and chargers. As with all aftermarket options, use them at your own risk.

If you do not use the batteries for an extended time, it is suggested they be stored with a 25-50% charge and not be allowed to drain completely.

BLACK LEVEL (Optional / Custom Mode only)

This is an adjustment that is only available in Custom Mode and is used to soften or deepen the appearance of black in your image. The setting can be adjusted from -15 to +15. Increasing the value lightens black and creates a more diffuse and soft image. Decreasing the value darkens black and creates a stronger image with more contrast. For some FS7 users, the default black level of the camera is too high, and it is recommended that this setting be lowered to -2 to -5. A recommended choice is -3.

There is another setting called Black Gamma on the camera, and it lets you change the shape of the gamma curve in the shadows. This is essentially a contrast setting for the shadow areas of your image. Changing Black Gamma from its default is not recommended and, in many instances, it is not even an option without conflicting with other settings. For instance, if you are shooting with any of the hypergammas or in S-Log, Black Gamma cannot be enabled. Furthermore, if Knee Saturation is enabled, Black Gamma cannot be enabled.

Be aware that when you dial the adjustment control for this setting, it requires additional turns of the wheel to register a change. If do not know this, it might seem like the control isn't working because you are turning the wheel, but the numbers aren't immediately responding.

> *That is why my advice to anyone who wants to make films is to make them, even if they don't know how. They'll learn by doing it, in the most vibrant and organic way possible.*
>
> **PEDRO ALMODOVAR**

CALIBRATION

It is important to do an initial and periodic calibration of your viewfinder. For this calibration, only the brightness and contrast are being adjusted.

To perform the calibration, go to:
CAMERA > COLOR BARS and turn the setting on.

Then under Type, choose SMPTE.

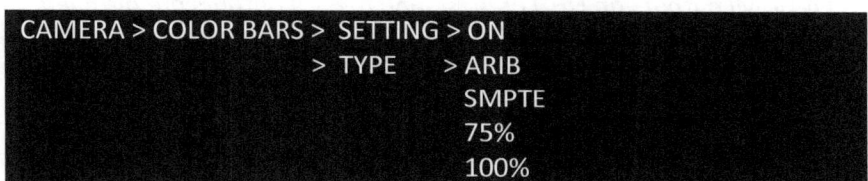

Once the color bars are displayed, adjust the Contrast control from the side of the Viewfinder to the exact point at which the white square first reaches maximum whiteness. If you have turned the contrast up all the way, then back it off until you notice the slightest decrease in whiteness.

Then go to:
VF > BRIGHTNESS > -99 to 99

... and adjust the brightness so that the -3.5 IRE bar (the bar on the left) disappears and matches the 0 IRE bar (the middle bar), and the 3.5 IRE bar (the bar on the right) is barely noticeable.

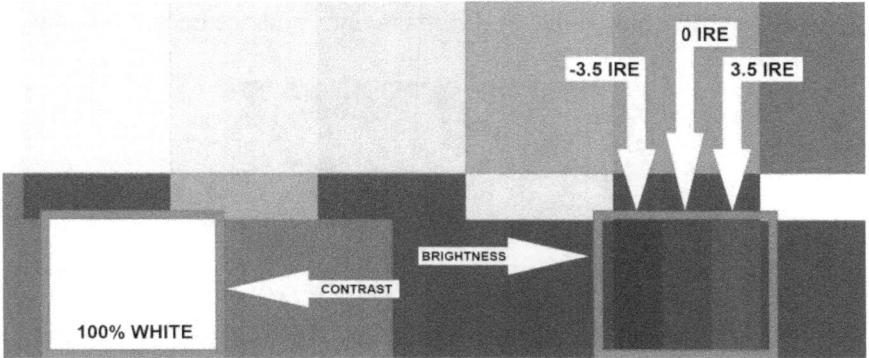

CAMERA OUTPUTS

Understanding the external camera output options only matters if you plan to use an additional monitor or recorder. There are a number of possible output options but to a large extent, the available options depend on which one of the 11 recording formats you have chosen. The Sony manual (pgs. 105-117) reviews all the possible combinations.

EXTERNAL VIDEO OUTPUTS (limitations)
1) SDI 1 and 2 have no UHD or 4K signal output.
2) If HDMI outputs an UHD or 4K signal, the SDI outputs are disabled.
3) SDI 1 cannot output text or overlays. (SDI 2 can if set in menu.)

VIDEO OUTPUT OPTIONS

		SDI 1	SDI 2	HDMI
High-Definition	(1920 x 1080)	Yes*	Yes*	Yes
2K	(2048 x 1080)	Yes*	Yes*	Yes
Ultra-High Definition	(3840 x 2160)	No	No	Yes
4K	(4096 x 2160)	No	No	Yes

* See #2 limitation above. There is no SDI output at all if HDMI outputs UHD or 4K.
NOTE: SDI 1 and 2 output a 10-bit 4:2:2 signal. HDMI outputs a 10-bit 4:2:2 signal up to 30 frames per second; above 30 fps, it drops to 8-bit 4:2:0.

CHANNEL 1 & 2 AUDIO-LEVEL CONTROL

The AUTO/MAN switch determines whether the audio-input levels are controlled automatically or manually. This is one of the rare settings that is controlled externally and cannot be adjusted from within the menu system. It is recommended that you always control your audio-input levels manually, so MANUAL is the position this setting should be in. This is a setting you should only have to make once and not be bothered with again.

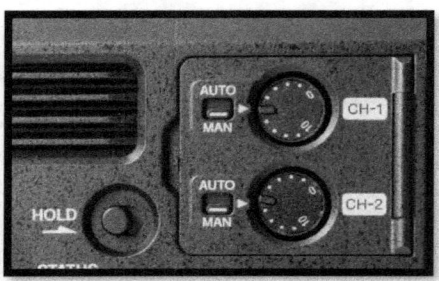

CINE-EI MODE

When the camera is set to CINE-EI mode, the camera, for most practical purposes, becomes a film camera. This means that like film, the digital recording process is locked at its native ISO and cannot be changed. You can adjust exposure by way of shutter speed, aperture, faster lenses, and external lighting, but ISO/GAIN is not an option. And through the use of a Monitor LUT, you can get a rough idea of what the footage might look like. And remember, you need to use the High/Low-Key functionality to really see into the upper and lower reaches of the image. And like a film camera, you can't really appreciate what you have until you are done filming and develop the recording by color correcting it later.

CINE-EI MODE

ENABLED FEATURES		
Exposure Index	Monitor LUT	High / Low Key
RAW Video*	White Balance Presets	
DISABLED FEATURES		
Auto White Balance	Auto Gain	Auto Shutter
Auto Iris	Auto Exposure	ISO / Gain
Paint Menu	Scene Files	Lens Files

* with XDCA-FS7, HXR-IFR5, and AXS-R5 attached.

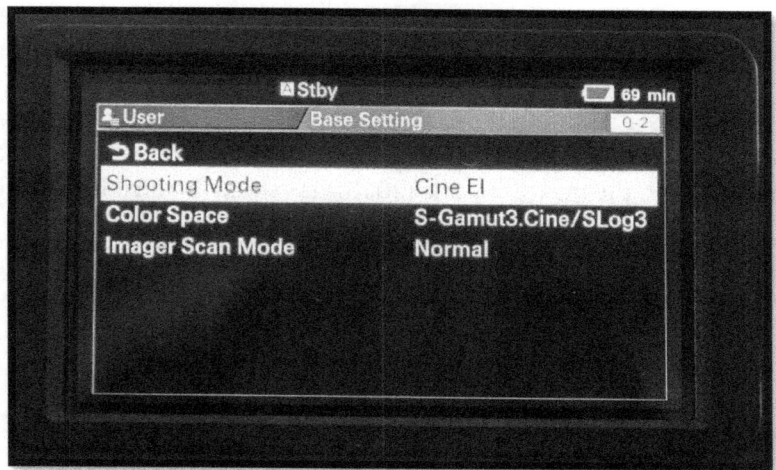

CLOCK SET

The clock settings are found under System > Clock Set.

Under these settings, you can set the time zone, date mode, 12/24-hour setting, date, and time down to hour/minutes/seconds.

CODEC

If your FS7 is not equipped with the additional XDCA-FS7 extension unit, you will find your choice of options limited. Your choices will be: XAVC-I, XAVC-L, and MPEG HD 422.

Of these options, XAVC-I is recommended. This is a robust, 4:2:2, 10-bit, intra-frame CODEC. This means each frame is individually compressed and does not rely on other frames for compression and decompression. This offers better results when editing and rendering.

In order to record ProRes codec, you would need to purchase the XDCA-FS7 extension unit. In 2018, these units are selling for close to $2,000 and give you the ability to record ProRes internally. Several popular external recorders can also record in ProRes, so that is an option worth exploring if you want your footage recorded in that format.

In a Sony-accessory world, if you wish to shoot in RAW, you would also need to purchase the HXR-IFR5 (interface unit) and the AXS-R5 (external recorder). In 2018, these two units combined cost close to what the camera does, approximately $7,500.

The good news is shooting in S-LOG can create imagery that closely rivals RAW footage, so you don't need RAW to create theatrical-release imagery. In fact, the movie Unsane (2018) was shot with an iPhone 7 Plus.

> *In any case, I tend to rely on only two kinds of lenses to compose my frames: very wide angle and extreme telephoto. I use the wide angle because when I want to see something, I want to see it completely with the most detail possible. As for the telephoto, I use it for close-ups because I find it creates a "real" encounter with the actor. Anything in between is of no interest to me.*
>
> **JOHN WOO**

COLOR SPACE (CINE-EI Mode only)

COLOR SPACE >	S-GAMUT / S-LOG2
	S-GAMUT3.CINE /S-LOG3
	S-GAMUT3/S-LOG3

The color-space option is only available if you are shooting in CINE-EI mode. To simplify the menu options, Sony has tied each color space to a single gamma curve.

The three combinations are:

S-Gamut / S-Log2
S-Gamut is Sony's original and proprietary color space. Its color-space triangle is the same size as S-Gamut3, and it provides a wide range of colors that exceed emerging standards such as the Rec. 2020 standard, which is the color-space standard for ultra-high-definition televisions. S-Log2 is a gamma curve that handles highlights better than S-Log3, clips at 107 IRE, and overall, is best for shooting in daylight or under lighting conditions you cannot directly control.

S-Gamut3.Cine /S-Log3
S-Gamut3.Cine is a smaller color space than S-Gamut, but still larger than the DCI-P3 space (the standard color space for digital movie projection) and much wider than the REC-709 color space for high-definition televisions. S-Gamut3.Cine is considered by many to give natural-looking color reproduction, mimics the characteristics of scanned film, and is designed to be easier to grade. S-Log3 is a gamma curve that handles low light better than S-Log2, clips at IRE 93, and overall, is best for shooting in low light or in controlled lighting situations.

S-Gamut3/S-Log3
S-Gamut3 is very close to the color space of the FS7's sensor and is optimized for color conversion when compared to S-Gamut. S-Gamut3 is considered suitable for shooting archival footage but might also prove to be more challenging to color correct. And as noted above, S-Log3 works best in low light or controlled lighting situations.

Whichever combination you choose, it is possible to grade any of them to look virtually identical.

COUNTRY (NTSC / PAL)

NTSC is the encoding standard that was used for broadcast television and DVD players in North America, Japan, and parts of South America.

PAL is the standard that was used for broadcast television and DVD players in Europe and many other parts of the world.

The default setting of the camera is usually the one you want to use as it will be set to the standard of the country where it is sold.

If that's not the case, then the setting you choose should be governed by where your intended audience is located.

As a rule, if the footage is destined for North America, Japan, and parts of South America, then you will want to select NTSC. If it is destined for almost anywhere else, you will probably want to choose PAL.

It is worth noting that while many PAL DVD players can also play NTSC video, most NTSC DVD players cannot play PAL source material.

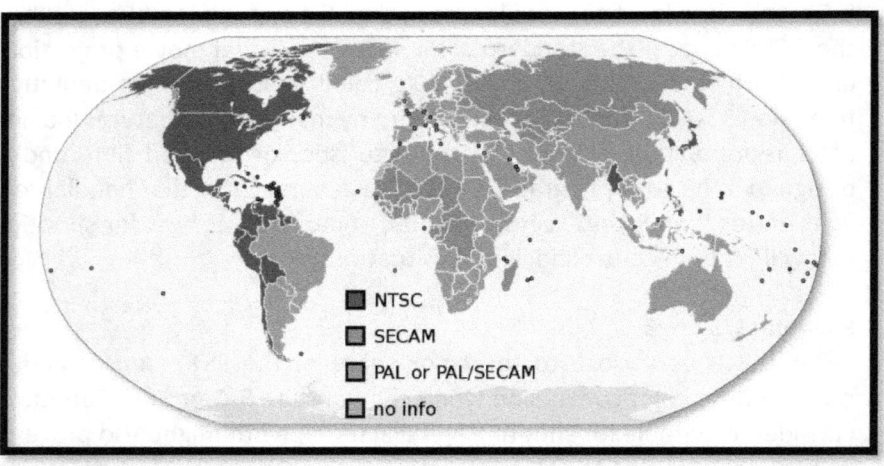

In other words, a film is a dream, but it's a dream made up of reality.
CLAUDE SAUTET

CUSTOM MODE

When the camera is set to Custom mode, it is a full-featured digital camera with myriad customizations possible. Custom Mode gives you access to nearly every feature the camera has except those reserved specifically for CINE-EI mode.

In Custom mode, you even have access to the S-Log curves that are really designed for CINE-EI mode.

Many auto settings become available in Custom mode, but it is never recommended that you use automatic settings. The problem is the camera can start making adjustments in the middle of a clip, and this can be visually distracting and problematic to fix later.

CUSTOM MODE		
DISABLED FEATURES		
Exposure Index	Monitor LUT	High / Low Key
RAW Video		

One of the primary advantages of Custom mode is the ability to have footage that does not require post processing and can be quickly edited and delivered. However, if you use one of the S-Log gamma curves in Custom mode, then you will have to grade the footage later either by hand or by applying a LUT.

About the only real disadvantage of Custom mode is that you are limited to hybrid REC-709 color spaces. But even with that, you still have a choice of five color schemes. Refer to page 74 for more information.

> *Of course, you may find that the biggest problem of young filmmakers is that they have nothing to say. At the risk of sounding redundant, I think the duty of the filmmaker is to tell the story that he or she wants to tell, which means that you have to know what the hell you are talking about.*
>
> **MARTIN SCORSESE**

CUSTOM WHITE BALANCE (Custom Mode only)

Once the exposure is set, you can now set a Custom White balance.

PROCEDURE

1. Try to fill as much of the frame as you can with a white or gray card. Using the gray side of a *Lastolite* card is recommended. Make sure as much light as possible is falling on the card and that the card is not in shadows. You will get a message from the camera if it does not receive enough light.

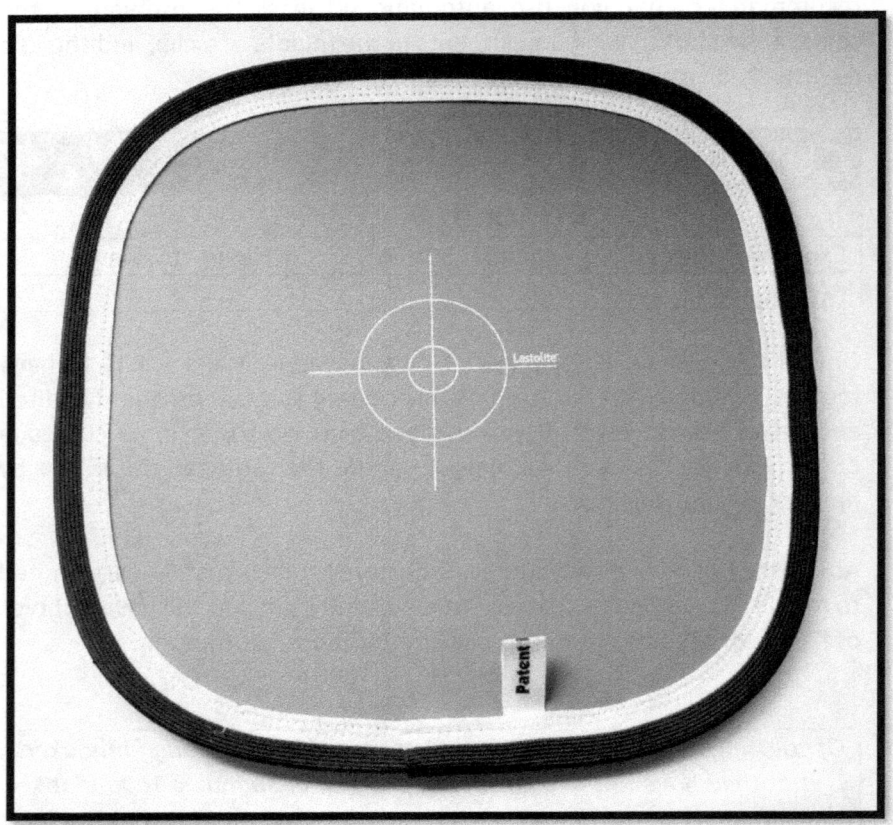

2. For the White Balance switch on the left side of the camera, make sure the switch is in either the A or B position. Position A and B each hold a single white-balance setting. Whether you choose A or B is up to you. The PRESET position holds a custom setting you can input through the Menu system.

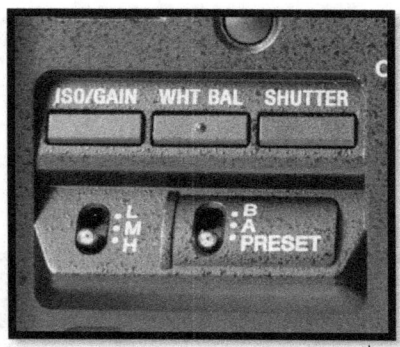

3. Once the card is properly positioned, press the WB SET button at the front of the camera. This will set a custom white balance for you and store it in either A or B memory.

Custom white balances are strongly recommended in mixed lighting situations.

White balance adjustments can be made in post but getting white balance right in the camera is strongly suggested and will provide better color reproduction.

The WHT BAL button at the side of the camera turns the ATW (auto tracing white balance) on or off. However, it is highly recommended that you never turn on the ATW. This setting will change your white balance settings as the lighting changes and can abruptly change the look of your footage while filming.

It is better to keep your white balance setting locked and only adjust it when the scene or location changes.

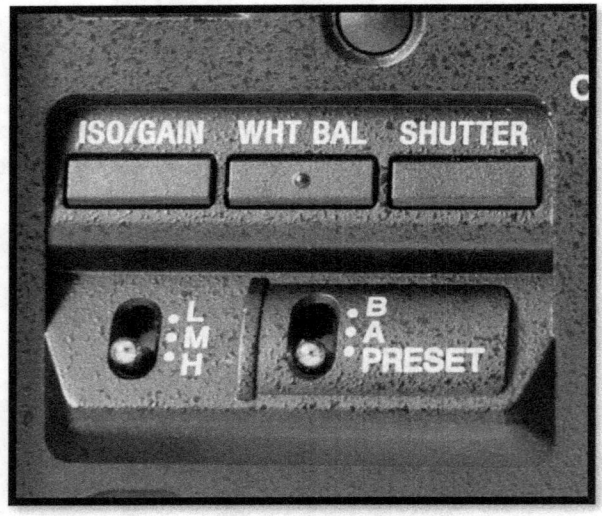

One good thing about the WHT BAL button is the center protrusion. This can help you locate the ISO/GAIN and SHUTTER buttons by touch. It is unfortunate that you cannot assign different functions to these buttons, but perhaps that might be changed in a future firmware update!

You need a proper exposure to get a custom white balance. This is why it is recommended that you set your exposure before white balancing. An error message will appear on the display if you try to get a white balance with insufficient light.

DISPLAY

Once you are at this point in the checklist, you should have most settings adjusted and be nearly ready to go. The only major adjustments left should be the exposure, white balance, and focus.

Once you have configured the majority of your settings, then double-check them by looking at the display. There are two ways to do this.

The first way is by looking at the viewfinder. It should reveal most key settings you have made and need to track.

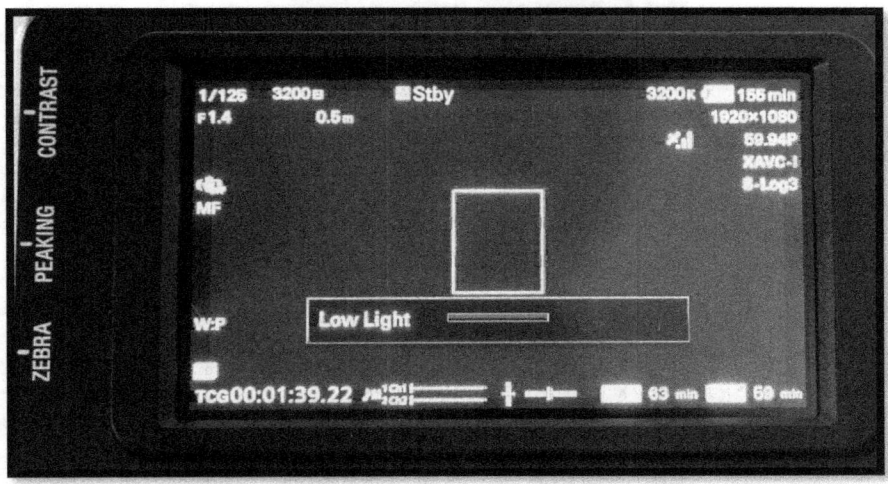

KEY SETTINGS DISPLAYED ON VIEWFINDER		
SHUTTER SPEED	ISO / GAIN / EI	WB KELVIN
BATTERY TIME	RECORDING FORMAT	FRAME RATE
CODEC	GAMMA CURVE	APERTURE
WB MEMORY SET	AUDIO LEVELS	AF/MAN FOCUS

There are more settings displayed than this, but these are the key ones. Under VF > Display On/Of, you can select the specific items you want on or off. Additionally, pressing the Display button on the side of the camera will turn the text display off and on.

A second way to check is by pressing the Status button, then using the Selection wheel to scroll through the various screens that display. In fact, you can check the status of every major system of the camera from the Status screens.

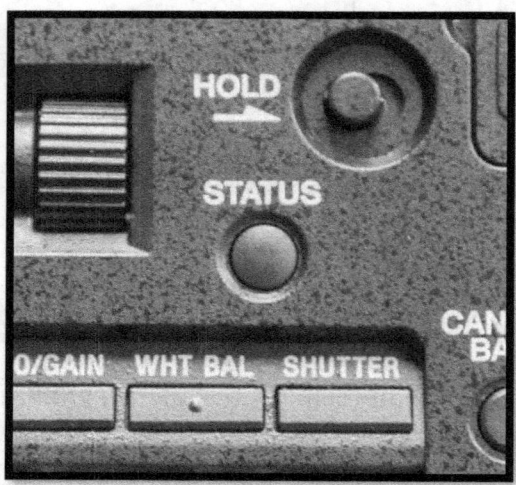

Once the Status screens display, pressing the Status button again turns them off.

EXPOSURE

Besides ensuring your subject is in focus, getting a good exposure is probably the most important thing you need to get right. Many shooters struggle with the FS7 simply because they don't get one.

There are several ways of getting a good exposure. The method reviewed here is just one of those ways, but it has proven itself to be efficient and effective. Hopefully, it will also serve you well.

This suggested method requires a 90%-white card. In the reference section of the book, there is a recommendation for a professional card you can consider purchasing.

You can use a piece of white paper instead, but note if you do that, you need to be aware that white paper is typically 2-4% brighter than a 90%-white card, and you may want to adjust your exposure levels accordingly.

COMPENSATION FOR WHITE PAPER			(% IRE)
	18% GRAY	90% WHITE	92-94% (WHITE PAPER)
S-LOG2	47	74	76-78
S-LOG3	56	76	78-80
STD-5	41	90	92-94

In general, using a piece of white paper may not be as accurate as using a 90%-white card, but it is much better than having no reference at all and can work quite well, especially with the appropriate amount of compensation applied. With S-Log, there is little, if any, harm caused by overcompensation; however, with the hypergammas, you do want to be careful about overcompensating and err on the side of underexposing.

MAIN POINTS

● Never underexpose S-LOG footage. As a rule, you will usually want to overexpose it by 1-2 stops. Usually the only time you would expose S-LOG to Sony's recommendation is under very bright light with few, if any, shadows.

● Never overexpose hypergamma curves. As a rule, you want to underexpose by 1/2 – 1 stop.

● Standard gammas, that is, REC 709, can be exposed to specs.

● In CINE-Mode you must turn off the MLUT to directly assess S-LOG footage with zebras; otherwise, you will be assessing the MLUT itself.

SUGGESTED EXPOSURE LEVELS FOR BETTER SIGNAL-TO-NOISE RATIO (%IRE)

	18% GRAY	90% WHITE	WHITE CLIP
S-LOG2	47	74	107
S-LOG3	56	76	93
REC-709 (MLUT)*	57	105	109

* This refers to the REC-709 (800%) MLUT if the MLUT and zebras are being used for exposure.

GAMMA EXPOSURE LEVELS (%IRE)

	18% GRAY	90% WHITE	WHITE CLIP
STD GAMMAS 1-6	41-42	90-96	100
HG1 3250G36	36	77	100
HG2 4600G30	30	68	100
HG3 3259G40	40	84	109
HG4 4609G30	33	74	109
HG7 8009G40	40	82	109
HG8 8009G33	33	73	109
S-LOG2	32	59	107
S-LOG3	41	61	93
REC-709 (800%) *	44	90	109

EXPOSURE (CUSTOM MODE)

PROCEDURE

1. Confirm which gamma curve you are shooting with. Make sure this is the curve you want and it is right for the lighting conditions.

2. Refer to the table on page 42 (or the back cover) and confirm what the proper IRE level is for 90% white.

3. If it is not already set, adjust your Zebra 1 value to the appropriate IRE value. The aperture can be set at 3%.

4. Put the white card in front of the camera and adjust the Iris or Aperture until you see zebra stripes filling the majority of the card.

You can also use a middle-gray card and use middle-gray values instead of white. However, for practical application, white is a common color to find and is easier to use for a reference. You should find the zebra function on the FS7 is accurate.

Although Custom mode is considered a what-you-see-is-what-you-get display on the viewfinder, you cannot rely on the display alone to determine the accuracy of your exposure.

> ### QUICK TIP
> *If you are in CINE-EI mode and toggle the ISO/Gain switch, but the image on the viewfinder does not change in brightness, then you know the Monitor LUT is off. It could also mean all three of your ISO settings are the same but that is unlikely.*

> *On the other hand, I tend to subscribe to the idea that, in a sense, you're always making the same film over and over again. You are who you are; your personality is usually the consequence of what you went through during your childhood, and you spend your life, consciously or not, rehashing the same ideas over and over again.*
> **TIM BURTON**

EXPOSURE (CINE-EI MODE)

PROCEDURE

1. Confirm which gamma curve you are shooting with. Make sure this is the curve you want, and it is appropriate for the lighting conditions.

2. Refer to the exposure table on page 42 (or the back cover) to see what the proper IRE level is for 90% white.

3. If it is not already set, adjust your Zebra 1 value to the appropriate IRE value. The aperture can be set at 3%.

4. Turn off the Monitor LUT.

5. Put the white card in front of the camera and adjust the Iris or Aperture until you see Zebra stripes filling the majority of the card.

6. Once the exposure is set, you can turn the Monitor LUT back on.

7. If the monitor is on, you can also keep an eye on where 90% white and near-white objects are falling on the wave form monitor. If the lighting changes or you get suspicious that something isn't right, you can easily turn the MLUT off and take another zebra reading. Be sure the waveform is monitoring the S-Log footage and not the MLUT. A title above the monitor indicates what is being monitored.

The reason you have to turn the Monitor LUT off is that the zebra function assesses the image on the viewfinder and not the underlying footage being recorded. If you don't turn the Monitor LUT off, then the Zebra function is assessing the exposure level of the displayed LUT and not the S-LOG footage.

What is a little counterintuitive is if you properly expose the REC-709 (800%) MLUT at 2,000 ISO, then you should also be properly exposing the underlying S-LOG footage. The relationship between them is fixed. If you hit 90% white accurately on the REC-709 (800%) MLUT at 2,000 ISO, then you are accurately hitting 90% white on the underlying S-Log.

SONY-RECOMMENDED
GAMMA EXPOSURE LEVELS (%IRE)

	18% GRAY	90% WHITE	WHITE CLIP
S-LOG2	32	59	107
S-LOG3	41	61	93
REC-709 (MLUT)*	42	90	109

* This refers to the REC-709 (800%) MLUT if the MLUT and zebras are being used for exposure. To directly assess the S-LOG footage, the MLUT must be turned off when you are taking zebra readings.

SUGGESTED EXPOSURE LEVELS FOR
BETTER SIGNAL-TO-NOISE RATIO (%IRE)

	18% GRAY	90% WHITE	WHITE CLIP
S-LOG2	47	74	107
S-LOG3	56	76	93
REC-709 (MLUT)	57	105	109

* This refers to the REC-709 (800%) MLUT if the MLUT and zebras are being used for exposure. To directly assess the S-LOG footage, the MLUT must be turned off when you are taking zebra readings.

The only issue with using the MLUT for exposure is that S-Log often requires an overexposure of 1-2 stops; therefore, in order to expose the S-Log footage properly you will also have to overexpose the MLUT by the same amount. You can use the zebra function for this since it will measure the exposure levels for whatever is displayed on the viewfinder, but you will have to increase the zebra levels to near maximum levels, for example, to 105 IRE.

You should experiment with different methods of getting exposure and see which one works best for you.

With the MLUT on in CINE-EI mode, it has been said that you can assess the exposure simply by looking at the viewfinder and guessing; however, that approach can lead to exposure assessments that are unreliable and inconsistent.

EXPOSURE INDEX AND EXPOSURE LATITUDE

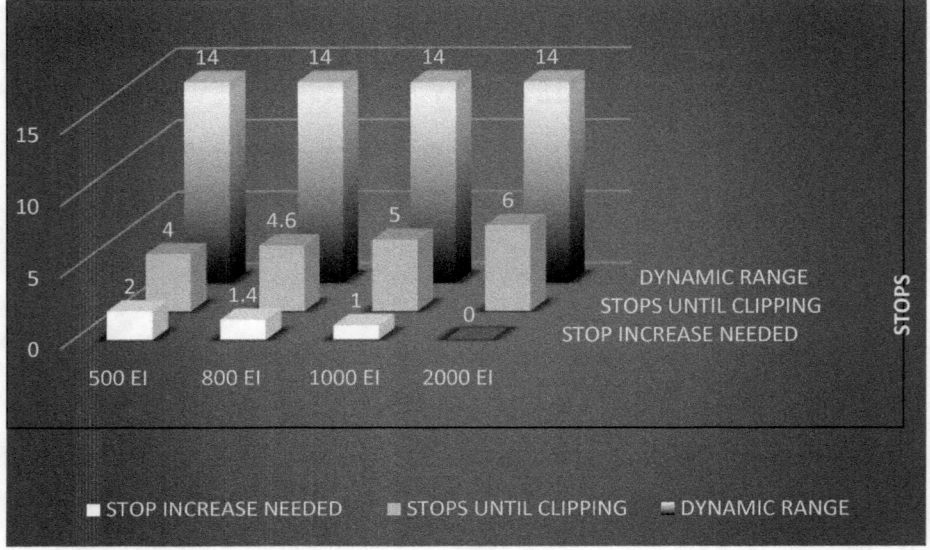

This chart illustrates the relationship between exposure latitude (i.e., headroom/stops until clipping) and possible exposure compensation in terms of stops (i.e., aperture). When the Exposure Index is set to 800 EI, then you will need to open up the aperture by 1.4 stops. Increasing the aperture in this way increases the signal, that is, the light hitting the sensor. This improves the camera's signal-to-noise ratio significantly. The only downside to doing this is it decreases the amount of headroom or exposure latitude you have until clipping occurs. If, for example, the EI is set to 800, then you would lose 1.4 stops of headroom. In most lighting situations, this is not going to be an issue. But in bright settings with few shadows and shaded areas, you might want to preserve as much headroom as possible to avoid blowing out the highlights. In a case like that, you might want to keep the EI at 2000 and not push the exposure beyond Sony's recommendation.

Please note that in CINE-EI mode changing the Exposure Index does not change the camera's native ISO of 2000 and does not alter the 14 stops of dynamic range. It only encourages an increase in aperture at the expense of exposure latitude and usually that is a good thing.

When shooting, it is advisable to keep a white object or 90%-white card nearby, so the camera's waveform monitor or an external waveform monitor can be used to make sure 90% white is staying near the correct IRE level (e.g., 74-76 IRE). A waveform monitor, even a tiny one like the one on the FS7's viewfinder, is a quick way of monitoring where 90% white is falling in terms of IRE. And as long as you keep your target white levels where they need to be, you will generally have good results in the end.

One thing that confounds the situation described above is Sony's own documentation, which recommends *standard* exposure levels that are below optimal for the best signal-to-noise ratio. For instance, Sony recommends that S-Log 2 be exposed for middle gray at 32%, and 90% white at 59%. Sony also suggests that S-Log 3 be exposed for middle gray at 41%, and 90% white at 61%. In almost all instances, everyone acknowledges these values are about a stop to a stop and a half too low for the cleanest possible image.

In a document titled "An Introduction to Log Shooting: What Creative Benefits does S-Log offer?" by Yoshihiro Enatsu and published by Sony, Enatsu explains:

"You can also ensure that noise is not noticeable by recording footage by brighter than standard, and then darkening it during grading. For example, you can shoot at 1 EV brighter than the standard S-Log2 exposure and then darken the footage at the same value during grading."

http://assets.pro.sony.eu/Web/ngp/pdf/an-introduction-to-log-shooting.pdf

Based on this and the experience of many others, it is safe to say the ideal signal-to-noise ratio for S-Log2 and 3 is achieved by overexposing by at least one stop beyond Sony's *standard* recommendations. See link for reference.

https://www.sony.jp/ls-camera/knowledge/pdf/TechnicalSummary_for_S-Gamut3Cine_S-Gamut3_S-Log3_V1_00.pdf

In summary, for the best possible imagery in most situations, overexposing S-LOG by approximately 1.5 stops is a good practice. The only time you might expose to Sony's recommendation is in bright daylight with few, if any, shadows. In that type of situation, you will already have ample lighting and will want to preserve as much of your highlight latitude or headroom as possible. In all other situations, overexposing by 1-2 stops is strongly suggested.

FIRMWARE

As of April 2018, the current firmware update for the FS7 is Version 4.20.

You can check the version under:

SYSTEM > VERSION

PROCEDURE

1. Make sure you have AC power attached.
2. Make sure have the proper SD card: 2-32 GB.
3. Put it in the camera and format the card, then eject it.
4. Download the latest firmware from Sony's website to your computer.
5. Unzip the file and transfer to the SD card.
6. Install SD card into the camera.
7. Then go to: **SYSTEM > VERSION**
8. On the menu choose:
   VERSION UP
9. You will be prompted to *Upgrade* to the newest version of the firmware.
10. Select *Execute*.

FIRMWARE UPDATE
https://pro.sony/ue_US/support/software/pxw-fs7-firmware-v420

FLICKER

```
CAMERA > FLICKER REDUCE > MODE > AUTO
                                 ON
                                 OFF

                     > FREQUENCY > 50 Hz
                                   60 Hz
```

If a situation arises and you are having issues with flickering lights, you can enable the Flicker Reduce setting. By default, this setting is off.

FOCUS

Getting a sharp focus is probably the single most important thing you need to do as a camera operator. If something is out of focus, it usually cannot be fixed in post processing. In addition, most people are sensitive to whether something is out of focus and will be distracted by it, especially if it is supposed to be in focus.

One of the tools the FS7 has for focus is peaking. If you use focus peaking and are comfortable with it, then you can use it to adjust focus.

The Focus Magnifier x4/x8 is also an excellent tool for checking critical focus, and if you use it, it is recommended that you assign it to Button 3. If you assign it, then by pressing the User menu button, #3, one time, you will automatically zoom in on your subject. And if you press it again, you can zoom in even closer. At this level of magnification, you should be able to achieve critical focus on your subject.

You will notice when you zoom in, you have the option of moving the area that is being magnified. If you use the Select wheel, you can move up or down, and if you use the button on either side of the wheel, you can move left or right. The magnifier can help ensure focus every time.

> *I'd rather see a well-composed static shot than a moving shot that makes no sense.*
>
> **JOE SIMON**

FOCUS ASSIST

Another useful guide when focusing manually is the Focus Assist Indicator. If you put your subject or item of interest inside the focus area, you will see that the length of the bar increases as the image sharpens. At a certain point, the bar will reach a maximum length and at that point, you should be in focus.

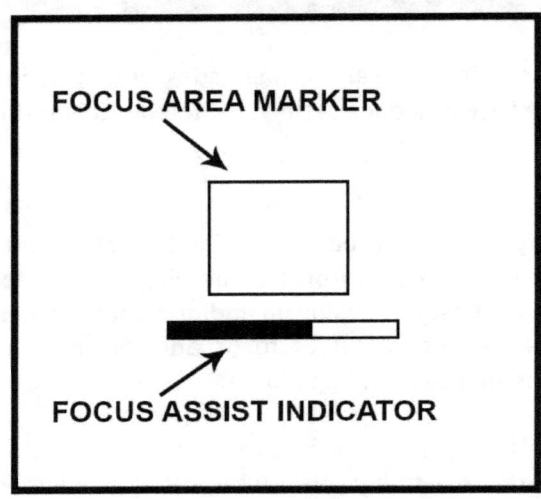

The bar never seems to reach the end; it only reaches a maximum length inside the indicator. For this reason, you usually have to focus in and out a couple of times to make sure you have reached the maximum length of the bar.

FORMAT MEDIA

Like the CODEC and Assignable Button options, this is another setting that appears more than once on the Menu system.

This is a straightforward function.

There are three possible cards you can format through this function; Media A, Media B, and the SD utility card. Simply select the card you wish to format, then execute the function.

GAMMA (Custom Mode only)

Next to Base Shooting mode, this is probably the second most important setting on the camera. There are some factors that will determine your decision.

First, the shooting mode you choose will affect the options available. If you decide to shoot in CINE-EI, the gamma setting will be disabled because in CINE-EI, the gamma curve is tied to the color space. If, however, you choose Custom mode, then you will have every gamma curve available to you. In some ways, this is great, but in others, it can seem like a daunting array of choices.

Second, you need to consider the lighting situation in which you will be filming. Is it a controlled or uncontrolled situation? Is it day or night? Is it a high or low-contrast situation? These factors should influence which curve you select.

Third, if you shoot with a S-Log gamma curve, then you will be looking at the need for grading in post, and at that point, you might be asking yourself: Why not just shoot in CINE-EI if it has to be graded anyway? Some might say that shooting in S-Log in Custom mode gives them increased control over their camera (for instance, being able to increase ISO or do a custom white balance) while giving them the advantage of a wider dynamic range. But the truth is the imagery will still need to have a LUT applied in post or be graded nearly the same as it would be in CINE-EI. For this reason, shooting with S-Log2 or 3 in Custom mode is not usually recommended.

One of the deciding factors to shoot in Custom mode is to avoid grading or applying LUTS in post, but if you shoot in S-Log 2 or 3 in Custom mode, then you will have to grade and apply LUTS, so why not just shoot in CINE-EI? If you are going to tailor the final look in post, then that *is* what CINE-EI is made for.

> But what you really should be thinking about is how to do a scene as efficiently as possible, as quickly as possible, and as intelligently as possible. Much of what people think of as art is actually just the intelligent solution of technical problems.
>
> **MICHAEL CHAPMAN**

GAMMAS AVAILABLE IN CUSTOM MODE

With the FS7, you have fourteen gamma options. At first, this might seem like being in a candy store, but once you look closer, your choices narrow quickly to three or four at most.

STANDARD GAMMAS

Under this category, you have six gammas available, but the only gamma you will probably ever want to use is STD 5, which is not only the most common standard or normal gamma but is also the one which most closely follows the REC-709 standard, the standard for high-definition television.

HYPERGAMMAS

There are six choices available. All represent a compromise in terms of factors such as clipping point, dynamic range, and curve design, all of which will affect how the curve responds to shadows, mid-range tones, and highlights.

S-LOG

Shooting in S-LOG in Custom mode is not suggested, but nevertheless, the FS7 gives you that option.

To reiterate, one of the primary reasons to shoot in Custom mode is to avoid grading in post, so if you shoot with a S-Log gamma curve, then the footage will require, at a minimum, the application of a LUT in post, if not more extensive grading work. In addition, in Custom mode, you cannot enable a Monitor LUT, and S-LOG footage can be difficult to see when filming.

If you are going to be applying LUTs in post and doing additional grading, then you might as well shoot in CINE-EI mode and give yourself complete control of the look you want to create—not to mention the ability to use a Monitor LUT when filming.

With that said, some filmmakers have no issues with shooting in S-Log in Custom mode and have no problems with applying LUTs or doing additional color corrections in post. There is no right or wrong on this point; it is primarily a question of preference more than anything else.

SUMMARY

BROADCAST-READY / NO GRADING / READY-TO-GO
STD 5 / HG 1 / HG 2
Of all the gammas, these three give you the most what-you-see-is-what-you-get appearance straight out of the camera. They are broadcast-legal HDTV gammas and are suitable for quick editing and immediate broadcast. Compared to the other gamma curves, they have more contrast and lower ISO levels (HG3 and 4 also have a *lower* ISO level). Of these three gammas, STD 5 would be the best choice for low-light situations and HG 2 the best for daylight.

LOW-LIGHT SITUATIONS
HG 3 / HG 7 / S-LOG3
These gammas give the best results in controlled or low-light situations. If the scene is dark and there are few, if any, bright lights, then either HG 3 or S-LOG3 would work. If it is a dark scene with multiple sources of bright light, for example, torches burning, then HG7 would handle the highlights better than S-LOG3. If you shoot S-LOG3 in low-light situations, then the footage will have to be graded and noise reduction applied. The other gammas might also require noise reduction in post as well.

DAYLIGHT SITUATIONS
HG 4 / HG 8 / S-LOG2
These gammas would be the best choice for uncontrolled or daylight situations. These three gammas have wider dynamic ranges and can handle bright lights and clipping better than their counterparts. Again, S-Log will require grading in post and so might HG 8 but not to the same extent that S-Log will.

The most challenging situations are usually in low light because you don't necessarily need the dynamic range that S-Log provides and may not have enough light to get an adequate exposure. The most common solutions are to bring in additional lighting or move closer to any available light sources, use a faster lens, shoot in Custom mode with one of the hypergammas, or be prepared to do extensive noise-reduction in post.

You are encouraged to shoot test footage using the different settings to see for yourself how these gammas perform under various conditions.

QUICK RULES FOR EXPOSING GAMMA CURVES

STANDARD OR NORMAL GAMMAS
For these gammas, it is recommended that you expose them to their specific 18% gray or 90% white levels. Do not under or overexpose them.

HYPERGAMMAS
It is recommended that you do not overexpose them. In fact, it is better to underexpose them by a half to whole stop and bring them up a little in post if needed.

S-LOG
The rule for S-Log is the opposite of the rule for hypergammas.

For S-Log, whenever possible, you should try to overexpose or push the exposure up by a stop or two. The only exception to this is in bright daylight with minimal shade or shadows, and you do not want to clip or blow out the highlights.

At a minimum, all the curves should be exposed to their middle-gray target values or to their 90%-white target values. Since white is a more common color than 18% middle gray, it is often more convenient to expose to 90% white than to 18% middle gray. Whether you choose gray or white is a matter of personal preference.

Using the camera's zebra function is a quick and reliable way of getting an accurate exposure; however, there are other methods that can work just as well. For instance, some people prefer to use light meters and skip all the built-in tools entirely while others prefer dedicated waveform monitors such as the *Leader LV5333* or *Tektronix WFM5200*. Achieving proper exposure is a destination and how you get there is up to you.

EXPOSURE TIPS AT A GLANCE	
STD GAMMAS 1-6	*Expose to specifications.*
HYPERGAMMAS 1-8	*Underexpose by ½ to 1 full stop.*
S-LOG	*Overexpose by 1 to 2 full stops.*

HIGH / LOW KEY (CINE-EI Mode only)

This function only works in CINE-EI mode and only when a Monitor LUT or MLUT is active. The MLUT by itself allows you to see a range of approximately 10 stops but since S-Log is 14 stops, the MLUT does not allow you the ability to see the entire dynamic range of the recording. However, with the High/Low Key function, you can see into the higher and lower reaches of S-Log's dynamic range.

For this function to work, it has to be assigned to a button. A good button to consider for this is Button 2. By default, button 2 is assigned to the Auto Iris function, but that's probably a feature you don't need as much as you might the High/Low Key function. Once you assign the button, it works as follows:

HIGH / LOW KEY BUTTON	
Press once.	**HIGH KEY**: Makes the display darker to see highlights.
Press again.	**LOW KEY**: Makes the display brighter to see shadows.
Press again.	Exits function.

HOLD FUNCTION

Once everything is set to your satisfaction and depending on what the circumstances are, you might wish toggle the hold switch. This will lock down any external buttons and dials and prevent them from being accidentally pushed.

The Hold function will also disable the recording button unless you turn that feature of the function off.

To enable recording with the Hold function on, go to:

SYSTEM > HOLD SWITCH SETTING > w/ REC > OFF

...and turn it off. You can also turn it off for the hand grip too.

This will let you record while locking down other controls that might be inadvertently pushed or turned.

INTERVAL RECORDING

Interval recording controls the time-lapse functionality of the camera and is simple to setup.

PROCEDURE

1. RECORDING > INTERVAL RECORDING > SETTING > ON
First navigate to the Recording menu and turn it on.

2. INTERVAL TIME > 1sec to 24hour
Choose an interval time; as a rule, the faster something is moving, the faster the interval time should be.

SUGGESTED INTERVALS

1-3 seconds	Freeway traffic, busy street, crowds.
3-5 seconds	Clouds, setting or rising sun, airplane traffic.
10-30 seconds	Astronomical phenomena.
1-2 minutes	Growing or blossoming plants.
5-15 minutes	Construction projects.

3. NUMBER OF FRAMES > 1 to 12
The number of frames available will depend on your frame rate. If you are shooting at 30 frames per second or slower, you should choose 1 frame per second. If you are shooting at 60 frames per second, choose 2 frames per second. However, at 2 frames per second, the results might not be as smooth. One frame is the recommended setting.

You can ignore the pre-lighting setting unless you have the proprietary Sony LED Video Light attached in the hot shoe on the handle.

4. Press record.
The camera will render the clip in-camera, and you can play it back immediately to see what it looks like. Also, keep in mind that time-lapse takes time to record, so plan accordingly.

ISO / GAIN / EI

As with most cameras, you will usually get a cleaner image by keeping ISO as low as possible.

In the case of the FS7, your base or optimal level of ISO depends on which gamma curve you are using. In Custom mode, there are three base levels.

BASE ISOs	
STD 1-6	800
HG 1-4	800
HG 7-8	1600
S-Log 2 & 3	2000

Let's explore the ISO/GAIN/EI Menu in more detail.

To begin, you have the Mode setting. This allows you to set your gain values as either ISO or dB. Some people like to see the ISO values and some prefer dB. For this guide, ISO is the preferred setting.

In Custom mode, you can assign an ISO setting to one of three memory slots: low, medium, or high. Within each slot, you can apply an incremental ISO or dB value. The exact values available from low to high will depend on the particular gamma curve selected and its starting value. Despite having a range of options, it is highly recommended that you keep your ISO or dB at its base or optimal level and assign that value to the L or Low position.

Once values are assigned to these slots, you can easily change between low, medium, and high by toggling the switch on the left side of the camera. However, keeping ISO at its lowest possible level or L is highly recommended.

RELATIONSHIP BETWEEN ISO/GAIN AND F STOPS
FOR S-LOG IN CUSTOM MODE

ISO / GAIN	STOPS (-/+)	NOISE FACTOR
8,000 (12 dB)	+2	x 4 increase
4,000 (6 dB)	+1	x 2 increase
2,000 (0 dB)	0	--
1,000 (- 6 dB)	-1	--
800	-1.4	--
640	-1.7	--
500	-2	--
250	-3	--

In Custom mode, when you increase the ISO/Gain setting you are actually adding electronic amplification to the signal and while this does brighten the image, it adds digital noise. For this reason, increasing the video gain or ISO is not usually recommended.

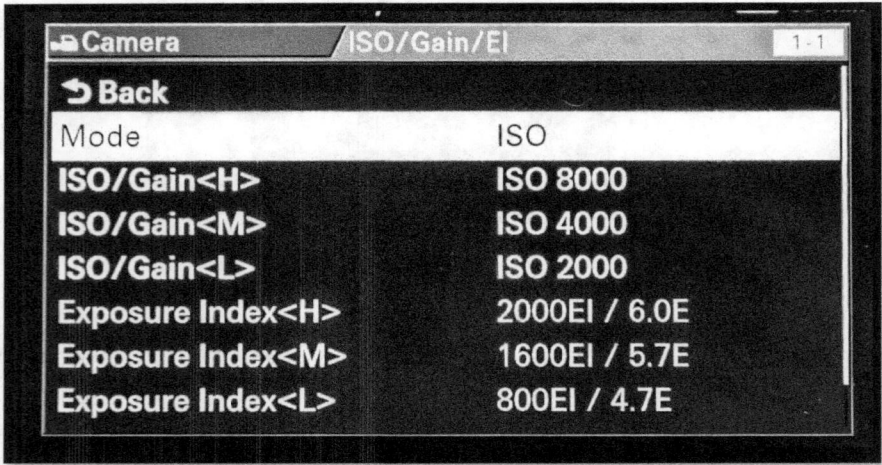

In CINE-EI mode, the camera is permanently locked at its native ISO of 2,000 and cannot be changed under any circumstances.

What can be changed is the Exposure Index (EI). However, the EI does not change the native S-Log ISO of 2,000 or alter its 14 stops of dynamic range.

The most noticeable thing the EI function does is precisely brighten or darken the Monitor LUT on the viewfinder, and this is supposed to visually cue you into making exposure adjustments. The EI function suggests what you need to do; it doesn't alter anything but the brightness of the MLUT itself. Now it is possible to burn the MLUT look into your footage but that is the only way a MLUT can *change* something.

The EI index only affects the MLUT, and this only affects the S-Log if you actually change the iris in response to what you are seeing on the MLUT. And if you do, for instance, open the aperture, then you are cutting into the headroom and pushing the exposure on the actual S-Log footage. This is why the EI number corresponds to another value, abbreviated with an E, that effectively shows how many stops you have before highlights start clipping.

Because the S-Log gamma curve has a significant amount of headroom or exposure latitude to begin with, in most situations, it is acceptable to cut into the headroom by overexposing 1-2 stops. By doing this, the light signal falling on the sensor increases and improves the signal-to-noise ratio. In post, you can pull the highlights down and end up with a cleaner image. You can overexpose S-Log because it has a wider dynamic range and more spare headroom than other gamma curves.

To see for yourself that the Exposure Index has no direct effect on the S-Log footage, all you have to do is put the camera in CINE-EI mode, then press *Record* and toggle the ISO/GAIN switch through L, M, and H. If you leave the camera's default settings on, this would be changing the EI from 800 to 1600 to 3200 ISO. It makes no difference if you have a MLUT on or off. When you review the footage, you will see changing the ISO/GAIN has no effect whatsoever. Now if you were to change the aperture that would cause a change but changing the ISO/GAIN in CINE-EI mode makes no difference at all.

You can think of the CINE-EI S-Log like film in a camera. It is what it is, and you can't change the film's ISO. Its properties are manufactured in. You can only change the basic camera settings such as aperture and shutter speed. EI works the same way. It can't alter the S-Log (film). It is only a tool for helping you visualize what changes you might want to make. But it is you who alters the S-Log through exposure adjustments, not the CINE-EI function.

In most lighting situations, which is to say interior locations, low-light situations, and shaded exterior locations, there is usually no harm in cutting into the exposure headroom in order to improve the signal-to-noise ratio. In those cases, there is not that much of interest in the highlights. The exception is on bright days with few, if any, shadows. In bright conditions, preventing clipping and not blowing out the highlights can be more important than trying to get a stronger light signal through overexposure. Under bright lights, you might want to expose to Sony's standard specifications and not push the exposure beyond that.

USER > ISO/GAIN/EI > EXPOSURE INDEX > H / M / L

In CINE-EI Mode, under the User menu and the ISO/GAIN/EI option, you can assign an Exposure Index setting to one of three memory slots: low, medium, or high. The options range from 500EI / 4.0 E to 8,000EI / 8.0E. However, when working with S-LOG, it is recommended that you never assign EI a value beyond 2000.

RECOMMENDED SETTINGS FOR CINE-EI
EXPOSURE INDEX SETTINGS

LOW	MEDIUM	HIGH
Option A		
500 EI / 4.0E	1,000 EI / 5.0E	2,000 EI / 6.0E
Option B		
800 EI / 4.7E	1,000 EI / 5.0E	2,000 EI / 6.0E
Option C		
640 EI / 4.3E	800 EI / 4.7E	2,000 EI / 6.0E
Option D (Default settings)		
800 EI / 4.7E	1,600 EI / 5.7E	3,200 EI / 6.7E

The numbers after the EI number refer to how many stops of headroom (aka exposure latitude) there is before clipping occurs. For example, 500

EI has 4 stops of exposure latitude while 1,600 EI has 5.7 stops. Option B is a good, middle-of-the-road choice.

Once exposure index values are assigned to these slots, you can easily change between low, medium, and high by toggling the switch on the left side of the camera. To reiterate, it is highly recommended that you keep the ISO/GAIN/EI switch at the lowest setting, which is L.

If you use the Option B settings and set the switch to L, this would put the camera at 800 EI and should serve you well in most lighting situations. There are, of course, exceptions to this but generally speaking 1.4 stops is a good overexposure setting for S-Log. In fact, 800 EI at the low or L level is the default Sony setting. It has been suggested that this was done by Sony because 800 EI is a good compromise between achieving an improved signal-to-noise ratio and preserving headroom.

And remember, EI only works when the Monitor LUT is on. If the MLUT is off, you will notice that toggling the ISO/GAIN switch of the camera has no noticeable effect on anything including the video recording.

To turn quickly initialize the MLUT, you would go to:

USER > MONITOR LUT > LUT Select

and choose:

P1: 709 (800%)

Then you would scroll down to:

USER > MONITOR LUT > VIEWFINDER > MLUT ON

And turn the MLUT on.

Once the MLUT is on and your EI values are set, if you go back to your display screen, you will see that toggling your ISO switch now darkens or lightens the image.

> *The only way you can make films for an audience is to make them for yourself. You have to use yourself as a reference.*
>
> **SYDNEY POLLACK**

KEY POINTS

● In CINE-EI mode, the camera records a S-Log gamma curve, either S-Log2 or S-Log3.

● In CINE-EI only, the ISO is locked at 2,000 and cannot be changed. It is the digital equivalent of film.

● S-Log has a dynamic range of 14 stops and will lack clarity if displayed on a monitor without a MLUT applied.

● Overexposing S-Log by one to two stops helps achieve a better signal-to-noise ratio and results in a cleaner image when corrected in post. However, this comes at the cost of decreased headroom.

● The Exposure Index precisely calibrates and controls the brightness of the MLUT, which then serves as a visual aid for suggesting exposure adjustments.

● The zebra function assesses whatever is displayed on the viewfinder.

● If the REC-709 (800%) MLUT is exposed to Sony's specification, then the underlying S-LOG footage will also be exposed to Sony's specification.
(Note: This assumes the ISO, shutter speed, and aperture are set the same for both.)

● The zebra function can monitor the exposure levels of S-Log footage directly, but the Monitor LUT (MLUT) must be turned off.

● All CINE-EI footage must be color corrected and cannot be properly judged until a LUT is applied or it is correctly graded in post. Uncorrected log footage is not meant to be seen without being graded or transformed with a LUT.

● In CINE-EI mode, you should never set an Exposure Index beyond 2,000. Or in other words, you should never underexpose S-LOG footage. It will contain excessive noise that will be virtually impossible to remove in post, even with noise-reduction software.

● Set the WB Preset as close to the lighting situation as you can.

ISO/GAIN/EI SWITCH

Before you adjust the exposure, it is a good idea to check that the external ISO switch is set to L. You can also double-check on the viewfinder and confirm that the ISO is set to the appropriate base level. Keeping ISO at its lowest setting, which should be **L** if it is properly adjusted, is highly recommended.

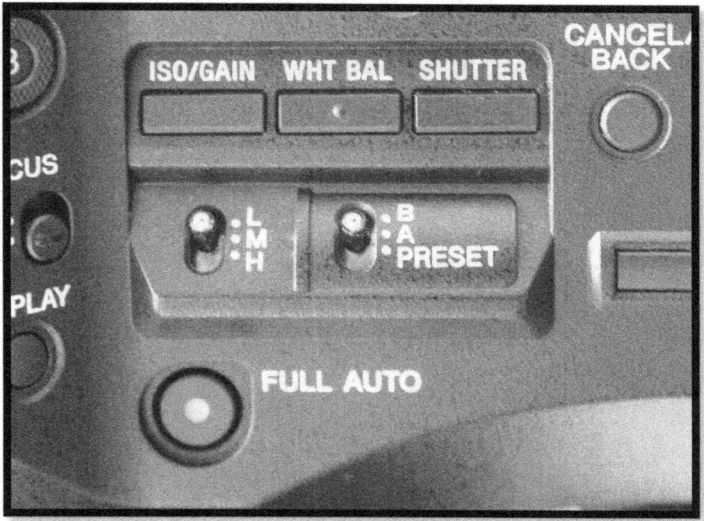

> **QUICK TIP**
> *The FS7 is future-proofed. It can shoot in S-Gamut3 color space, which is larger than Rec. 2020, the standard color space for UHDTV. If you want to deliver a project that meets the Rec. 2020 standard, then shoot in S-Gamut3 and apply a LUT that transforms it to Rec. 2020.*

> *Whether it's the right way to construct things or not, it's personal. This is my own reaction to seeing too many things that were strung together on illusion. It's sort of a corollary to trying to keep the space is that you don't cut. That you have as few cuts as possible. Kubrick did, I think, that there are fewer than 500 cuts in Clockwork Orange.*
> **JOHN McTIERNAN**

LANGUAGE

The language setting is found under **SYSTEM > LANGUAGE**.

If the camera is not already set to your language of choice, it can be easily set to your preference.

LANGUAGES AVAILABLE IN THE FS-7		
English	French	Spanish
Mandarin	Cantonese	Portuguese

LENS

Once the lens cap is removed, double-check the lens to make sure it is clean and free from dust, water spots, and smudges. On a side note, if you plan to shoot in S-Log, you should consider investing in and using the fastest lenses you can.

LENS CAP

For safety and other reasons (e.g., running the APR function), it is recommended that you leave the lens cap on until you are ready to start adjusting your exposure, white balance, and focus. When you are ready, then remove the lens cap and remember where you put it!!

LENS HOOD

Once the lens is confirmed clean, you should install your lens hood or matte box. It is recommended that you never shoot without a lens hood installed. A lens hood will not only reduce unwanted lens flare, it will help keep the lens clean and protect it from getting accidentally scratched or damaged.

> *In terms of visual grammar, for me, the spatial relationship between the characters is the vital thing. If characters are emotionally close, I bring them close; if they're distant, I separate them.*
>
> **JOHN BOORMAN**

LOOK PROFILES (CINE-EI Mode only)

Monitor Look Up Tables (MLUTS) are 1D LUTs and affect gamma, but not hue or saturation. Look Profiles are 3D LUTs or cubes that affect gamma, hue, and saturation. Given the degree to which 3D LUTs can manage the overall look and color of an image, they are used primarily for grading but can also be used for monitoring like 1D LUTs. Note: 1D LUTs can be used for grading purposes, but 3D LUTs are more commonly used.

MONITORING LUT OPTIONS

CATEGORY	TYPES
LUTS [1D LUTS]	**709 (800%)**: *Traditional REC 709 with improved dynamic range.*
	HG8009G40: *Good dynamic range, better for low light.*
	HG8009G33: *Good dynamic range, better for day light.*
	S-Log2: *Excellent dynamic range, better for day and uncontrolled lighting situations.*
	S-Log3: *Excellent dynamic range, better for low and controlled lighting situations.*
LOOK PROFILES [3D LUTS]	**LC-709**: *Low contrast, less saturation, meant to be graded.*
	LC-709 Type A: *Low contrast, less saturation, meant to mimic the look of an Arri Alexa camera, which means more saturation in the shadows and mid-tones than in the highlights.*
	Slog2-709: *Less saturated than REC-709, dynamic range of S-Log2, meant to be graded.*
	Cine+709: *Color film look, meant to be graded.*
USER 3D LUTS	User created 3D LUTs.

The primary purpose of a Look Profile is to give you an idea of what your footage will look like in post with a particular Look Profile (aka 3D LUT) applied. With a Monitor LUT, instead of waiting to see what the footage might look like once graded, you can see it while you are shooting.

It is important to remember that whatever 3D LUT you apply in post must match the S-Log and Color Space of the recorded footage. For instance, if you shoot in S-Log3 with the SGamut3.cine color space, then the 3D LUT you apply in post must be one that converts S-Log3/SGamut3.cine to 709. In Da Vinci Resolve, it comes with four Sony 3D LUTs (aka Look Profiles) already loaded in, but these LUTS only work if you record in S-Log3/SGamut3.cine.

If you were to shoot with S-Log2 in CINE-EI mode, then you could still use the Look Profiles, but would need to find the Sony Look Profiles online and download them into Resolve, so you could apply the right LUT to the right Log footage.

In grading footage with a LUT, the Look Profiles are usually starting points and are meant to be graded beyond what the LUT does to the image; however, it is possible that the LUT renders the image perfectly and little else will be needed. Some filmmakers like to create their own 3D LUTs and upload those profiles into the camera. Then they use those LUTs to monitor their footage and once they're done, they apply the 3D LUT to their footage in post.

You can also use the 3D LUTs to monitor your exposure but need to make sure that you map your LUT to the correct gray or white level. See the next chapter on Mapping LUTS.

BEWARE!

If you enable:

VIDEO > MONITOR LUT > SDI1 & INTERNAL REC > ON

...then you will burn the MLUT into the underlying S-LOG footage. For this reason, it is recommended that you keep this setting off unless you are certain this is what you want.

MAPPING LUTS (CINE-EI Mode only)

This book recommends getting an exposure by using the zebra function and assessing the S-Log footage directly without a MLUT. However, if you want to use zebras with MLUTs to assess your exposure, you need to know how each MLUT is mapped from the underlying S-Log curve.

For the procedure below, we will be using the Look Profile, 2.LC-709TypeA and SGamut3.Cine/Slog3. This procedure also requires a 90%-white card.

PROCEDURE

1. Switch to CINE-EI mode. Choose: Color Space: SGamut3.Cine/Slog3.
2. Go to: ISO/GAIN and set EI <H> to 2,000. Then make sure your ISO Switch is set to H, which should put it at 2,000.
3. Go to Video > Monitor LUT and turn *all* the MLUTS off.
4. Go to VF and turn the Waveform On and set the Source to SDI2.
5. Go to Zebra setting, turn On, set to Zebra 1, set level to 61%, and aperture to 2%. [note: 61% is Sony's *standard* recommendation. In actuality, we would recommend 76% as the ideal exposure level.]
6. Exit the menu. On the Display you should see the waveform monitor and the title above should say S-Log3. Confirm EI is 2,000.
7. Take a 90%-white card and adjust the aperture until the zebras show up on the card, indicating a 61% level. At this point, the exposure level for S-Log3 should be correctly set, that is, according to Sony.
8. Leave the white card and camera in its position.
9. Be sure not to adjust any of your settings from this point. That is, leave the ISO, shutter speed, and aperture alone.
10. Go back into the Menu. Go to: Video, Monitor LUT.
11. Go to: Category. Select Look Profile.
12. Go to: Look Profile Select and choose: 2.LC-709TypeA.
13. Go to: SDI 2 and turn the MLUT on.
14. Go to: Viewfinder and turn the MLUT on.
15. Exit the menu.
16. Looking at the display, you should see that the waveform monitor is still on, but now monitoring the MLUT profile.
17. With the white card still in front of the camera, note where the white card's level is falling in terms of IRE on the waveform monitor. This will only be an approximation.

If everything went correctly, the 90%-white card should be showing at or around 71% IRE on the waveform monitor.

CORRECTLY EXPOSED S-LOG WILL LEAD TO THE IRE VALUE FOR A CORRECTLY EXPOSED MLUT (%IRE)	
	90% WHITE
S-LOG3	61
LC-709 Type A	71

If you expose S-Log for 90% white (which in this example is 61% IRE), then turn on the Monitoring LUT (being sure to keep the same aperture, shutter speed, and ISO), the waveform monitor will reveal the IRE level needed for a proper exposure of the MLUT at 90% white, which, in this case, should be 71%. The key is making sure both the MLUT and S-Log keep the same aperture, shutter speed, and ISO settings. And remember, since S-Log's ISO is locked at 2,000, you must set the EI to 2,000 for the MLUT too; otherwise, it won't work or map properly.

Using the waveform monitor only provides an approximation because the display on the viewfinder is so small. To get a more accurate reading, you can also use the Zebra 1 function to confirm the target value. To do this, you would go to Video > Monitor LUT > Viewfinder and turn the MLUT on. Then within the Zebra 1 Level function, you would adjust the Zebra 1 Level and take note at what percentage zebras start dominating the 90% card, which, in this example, should be very close to 71%.

In summary, if you have a known-reference value such as a 90%-white card and can get a correct exposure for S-Log using the card, then when you switch to the monitoring LUT and look at the waveform monitor (or check the zebras), the 90% card will fall at a level on the waveform monitor that reveals the IRE value needed for a correct exposure of the MLUT at 90% white. In short, given the same settings, if the underlying S-Log is correctly exposed, then the MLUT will also be correctly exposed and vice versa.

> *When I'm working on a script with James Schamus, I always force him to come up with one word to sum up the essence of a scene or a movie—to sum up the core emotional feeling or taste of the movie.*
>
> **ANG LEE**

MATRIX (Custom Mode Only)

When you are shooting in Custom Mode, the FS7's color space will be locked in Matrix mode. This is a Sony-engineered color system based on the REC-709 HDTV color standard. Each of the menu options in this section is briefly reviewed below.

PAINT > MATRIX > SETTING > ON

This setting enables all the menu options beneath it. If you turn it off, none of the settings below it will take effect. Generally, it should be on.

PAINT > ADAPTIVE MATRIX > ON

According to Sony, this function helps control areas of an image that show up as excessively blue, especially under LED lights, which are disproportionately blue. Due to the prevalence of LED lighting, there is little harm in keeping this setting enabled.

PAINT > PRESET MATRIX > ON

In order to choose one of five color matrices, you must first turn this setting on. If it is left off, the Color Matrix will be left on the default mode: Standard.

REFER TO PAGE 74 FOR ADDITIONAL SETTINGS.

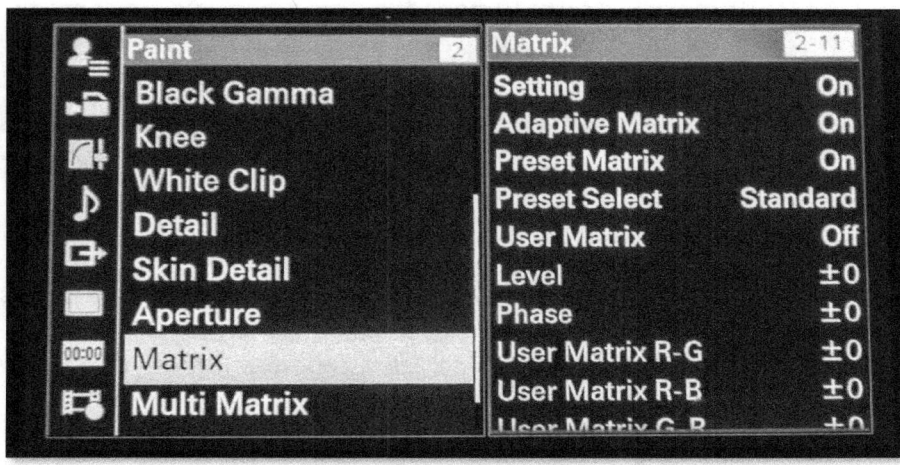

MONITOR LUT [MLUT] (CINE-EI Mode Only)

This setting can be activated through the User menu, but you can also activate it in the Video menu. Since there are other settings you will need to confirm or adjust along with the MLUT in the Video menu, it is recommended you wait until you get to the Video menu to set it.

However, when it comes time to turn off the MLUT quickly, as might be the case when you are taking zebra exposure readings, then you will want to turn the MLUT off and on from within the User menu.

By default, you first encounter the MONITOR LUT setting in the User menu. And again, by default, it will be set to:

LUT SELECT	P1 709 (800%)

It is recommended that you leave this setting as is, and it should be sufficient for most, if not all, circumstances.

USER > MONITOR LUT > VIEWFINDER > MLUT ON

This enables the use of the Monitor LUT.

With the Monitor LUT on, the image in the Viewfinder becomes a what-you-see-is-almost-what-you-get or *WYSIAWYG* image. The MLUT makes it easier to get a ballpark idea of what your image might look like if it were displayed on a standard HDTV monitor; however, since the REC 709 (800%) LUT can only display 10 stops of dynamic range, you can't see the low or high end of S-Log's 14-stop dynamic range. (See page 55.)

MLUT NEEDS THE *HIGH/LOW KEY FUNCTION* TO DISPLAY FULL RANGE		
LOW	709 (800%) = 10 STOPS OF RANGE	HIGH
	S-Log = 14 STOPS OF RANGE	

If the Monitor LUT is off, then you will see a washed-out image that is difficult to assess visually. The only time that the Monitor LUT should be off is when you want to check the zebra exposure levels on the S-Log footage directly.

MULTI-MATRIX (Custom Mode only)

PAINT > MULTI MATRIX > SETTING > ON

This setting enables the Multi-Matrix settings beneath it and allows precise control over the colors in your image. The Multi-Matrix allows a surgical approach to color control while the User Matrix affects colors interactively like touching a spider web.

The Multi-Matrix is also considered an advanced-user setting and like the Matrix settings, changing these settings requires experience and knowing exactly how to achieve your intended look.

Despite the importance of having experience and the right tools, there are two settings you can easily change with little experience. These suggested changes are based solely on what has been perceived by some as the FS7's bias of overemphasizing red *and* green-yellow.

TWO SUGGESTED MULTI-MATRIX ADJUSTMENTS
1. Select YL (Yellow) under Axis and decrease saturation to -25.
2. Select R (Red) under Axis and decrease saturation to -20.

If you decide to experiment with the multi-matrix settings, there are two ways you can select a color to adjust.

The first way is by enabling Area Indication, then selecting Color Detection. After you've done that, a tiny square will appear in the middle of the Viewfinder. Line that square up with the color you wish to identify and adjust. The camera will then determine which color axis matches, and the symbol representing that hue will appear. Using color detection may not always be 100% accurate, but at least it gets you in the ballpark.

> *I enjoy not doing the same thing. I like intimate movies like Last Tango in Paris or Besieged to have an epic flavor. In the same way, I like male actors to have something feminine about them and actresses to have something masculine.*
>
> **BERNARDO BERTOLUCCI**

The second method is to select one of the 16 colors by its letter code. Generally, it is a safe approach to stick to one of three primary hues, R(RED), G(GREEN), or B(BLUE), or one of the secondary colors, MG (MAGENTA), CY(CYAN), or YL (YELLOW). The ten remaining colors represent varying degrees of blends between these colors (tertiary and beyond). Each of the 16 colors occupy a slice or swath of the color wheel or circle. For instance, if you point the camera at a color wheel and decrease the saturation on any of the 16 colors to -99, the color effectively disappears and is left gray. Since multi-matrix settings offer a precise degree of control, it takes high values to fully saturate or desaturate a specific color.

It is also worth noting that *pure* colors rarely exist in nature and that most colors in the world are a blend. For instance, green often has yellow in it, so it is rare to see something that is 100% green. This means that decreasing what you perceive as excessive green in your image might actually mean you need to decrease YL+, YL-, or another blended hue.

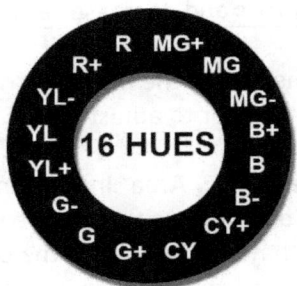

Unfortunately, only one-color axis can be displayed or changed at a time. And there is no summary page that displays the changes you have made; therefore, you have to take note of all the individual changes you make or be prepared to pull up every hue by itself to see if it has been altered.

This is one reason it is recommended that you reset a camera if someone else has used it before you. You don't know what setting might have been altered, and it is a time-consuming process to check everything.

NOISE SUPPRESSION

The FS7 has a noise-suppression feature that works in Custom or CINE-EI mode. By default, it is on and set to medium. Some FS7 users have taken the position that a properly exposed image that is graded well in post does not require in-camera noise suppression; however, Sony must have had a reason they added this feature and turned it on by default.

The reality is that in almost any low-light situation some noise is bound to turn up in some parts of the shadows no matter what you try to do. In fact, in some conditions, even with the aperture wide open, you might not be able to hit your recommended exposure levels, and noise will be unavoidable.

In low-light conditions, not only should noise suppression be on, it should be turned to high. Even if noise suppression is at its highest setting, in some situations, you might still need to use the noise-reduction features within a program such as Da Vinci Resolve.

Neat Video also sells an excellent noise-reduction program that installs in all major video editors; however, no matter which program you choose to use, noise reduction requires extensive computer processing and can take many hours to render. This is why it is important to get the best exposure levels possible and avoid the need for noise reduction in the first place.

PAINT (Custom Mode only)

```
PAINT > PRESET SELECT > STANDARD
                       HIGH SATURATION
                       FL LIGHT
                       CINEMA
                       F55 709 LIKE
```

With the Preset Matrix *(see page 69)* turned on, the Preset Select lets you actually choose from one of five specially-modified color systems.

STANDARD
The default setting, a REC-709 HDTV image; this is what some might call a *normal* image.

HIGH SATURATION
Boosts the color intensity beyond the Standard.

FL LIGHT
Helps compensates for the green cast in fluorescent lighting and provide a more natural appearance.

CINEMA
A slightly desaturated, softer look.

F55 709 LIKE
Compared to the FS7 Standard mode, F55 709 LIKE appears slightly more saturated with warmer/more natural skin tones. It emulates the color response of Sony's Cine Alta F55 camera.

I think Traffic works for people because it looks and feels like it's going on right in front of you, and that means shooting with available light with handheld cameras. The reason I'm drawn to this approach is that I feel that if you are going to do something that is lifelike, then it will be inherently self-renewing because life changes, and so if you keep the idea of trying to capture life as it happens, then your talent won't desert you, whereas with any other style, you know you'll run into a dead end.

STEVEN SODERBERGH

PAINT > USER MATRIX > ON

This setting enables the User Matrix, which lets you adjust the overall color scheme of your image.

Most settings under the User Matrix are advanced and making lots of adjustments here is not suggested unless you know how to paint the camera and have a correctly calibrated monitor, scopes, DSC charts, and a lot of time, patience, and practice. However, with that said, there is one adjustment that is safe for a beginner to make and that is the Level setting.

PAINT > LEVEL > -99 to 99

This setting adjusts the saturation or color intensity of the overall image. Increasing the number raises the intensity and decreasing it lowers the intensity. This is again a preference setting, and you should feel free to experiment with it if you want more colorful and vivid imagery. However, the default setting of 0 is acceptable to start with.

PAINT > PHASE > -99 to 99

This setting alters the actual hue of all the colors. If you can visualize the colors on a vectorscope or wheel, then positive values shift all the colors clockwise, and negative values shift everything counterclockwise. As you might imagine, this can have many unintended consequences and should be done only if you have the right tools and know what you are doing.

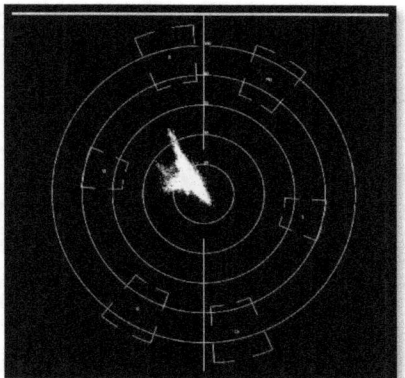

The distance a hue stretches from the center indicates the chrominance level. Achromatic values (black, gray, and white) show as single dots in the center of the scope and indicate no chroma or color information.

USER MATRIX (LINKED PAIRS)

KEY CONCEPTS FOR USER MATRIX					
R-G	R-B	G-R	G-B	B-R	B-G

Any changes affect the saturation of the first color and **the saturation and hue** of the second color.
A positive value (0 to +99) increases saturation for both colors but also alters the hue of the second color.
A negative value (0 to -99) decreases the saturation for both colors but also alters the hue of the second color.
A positive value adds the first color to the second; a negative value subtracts it.
A positive value shifts the second color clockwise; a negative value shifts it counterclockwise.
Any changes can affect the other colors in hue and saturation, especially colors neighboring the second color.

Making an adjustment to the User Matrix pairs is not recommended unless you have the tools and experience to adjust them. If you do have the tools and time, then these settings can be used to make in-camera color corrections, create custom looks, and match cameras; however, adjusting the overall saturation requires no special experience or tools.
-'
A case can be made that Sony's default level of saturation is too high for what might be considered a *film look*. If you are interested in that kind of a look, then you will want to consider dialing down the saturation level.

PEAKING (Optional)

This is an optional setting and depending upon your circumstances and personal style, you might want to enable this focusing tool.

```
USER > PEAKING > OFF
                 ON

> PEAKING TYPE > NORMAL
               > COLOR

      > COLOR > B & W
                RED
                YELLOW
                BLUE
```

If you set the peaking type to Normal, you won't have any color options, even though the color settings do not dim and still appear as options.

Peaking can be helpful when you need fast verification that your primary subject or scene is in focus. This could involve fast-moving situations or circumstances where you are having difficulty seeing the image clearly on the viewfinder for some reason.

The downside to peaking is that if you magnify the image to check critical focus, then the peaking colors obscure lines and edges, and you have to play with the focus to see where the peaking signal reaches its maximum, and even then, you still can't see the actual edges of the subject. If you don't play with the focus to achieve the maximum peaking signal, then your focus could be soft in some areas.

There is no right or wrong way to check focus, and a person's choice of focusing tools is a preference.

> To conclude, it's all really about storytelling. I think you have to know how to tell a story. It doesn't matter whether it's a movie or a joke over dinner. There are people who start telling a joke, and you cringe. Others open their mouths and you're hanging on their every word.
>
> **ROMAN POLANSKI**

POWER

The power switch is located on the left side of the camera, toward the bottom and near the back.

IT IS CRITICAL THAT YOU REMEMBER TO TURN THE CAMERA OFF BEFORE YOU STORE IT AWAY. IF NOT, THE CAMERA COULD OVERHEAT AND CAUSE SERIOUS DAMAGE. THE CAMERA HAS NO POWER-ON LIGHT.

If the switch is pushed toward the back, it is off. If the switch is pushed forward, that is, to the front of the camera, it is on.

There is no official power light for the camera, but if at least one XQD card is installed, then when the camera is on, the card indicator light should be green; therefore, although it is not an official power light, you can treat the card indicator light as a de-facto power light.

PRESET WHITE BALANCE (CINE-EI Mode Only)

When you are shooting in CINE-EI, you cannot perform a custom white balance and are limited to three presets. These are controlled by the external WB switch.

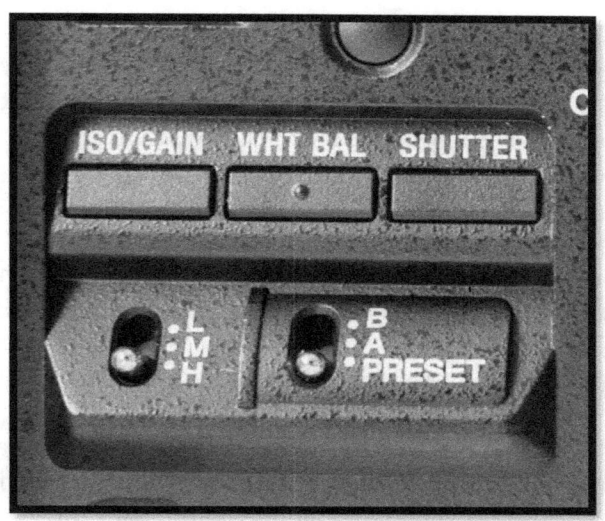

WB PRESETS (CINE-EI MODE)	
B	5500 K (daylight)
A	4300 K (mixed)
PRESET	3200 K (incandescent)

With regard to white balance, it is assumed you will try to get it as close as you can to the correct Kelvin number and will fine tune it in post. For this reason, it is usually advisable to get a shot of a white card in each location, so you can use that for reference when grading.

It is also advisable to match the white balance to your general setting. Having the incorrect setting for the situation can lead to an increase in visible noise.

For instance, if you have the white balance set to 3200 K (incandescent) when you are shooting outside on a bright day (5500 K or higher), then the blue channel is going to be unnecessarily amplified and might show noise. Blue is amplified under a 3200 K setting because incandescent light lacks blue; therefore, the white balance function will multiple the blue channel by a larger factor than the red or green channel.

REC FORMAT

The FS7 has eleven recording formats to choose from. The options range from 4096 x 2160 @ 59.94p to 1920 x 1080 @ 23.98p.

One suggestion is to consider is shooting your footage at 59.94 (that is, 60 frames per second) and dropping the footage into a 30 fps timeline on your video editor. If you do this, then the project will still render out at 30 fps and simply drop the frames it doesn't need; however, if you want to slow the footage at certain points, then you will have enough frames available to render smooth slow-motion footage at ½ speed.

Although the frames per second that you choose will appear on your display screen, it is good to take a mental note of your frames per second. Generally, your shutter speed will need to be double your frame rate.

AVAILABLE RECORDING FORMATS				
4K (4096 x 2160)	59.94p	29.97p	24p	23.98p
UHD (3840 x 2160)	59.94p	29.97p	---	23.98p
HD (1920 X 1080)	59.94p	59.94i	29.97p	23.98p

RECORD

If the arm grip is attached, the FS7 has three record buttons that can be activated—one on the grip, one on the top handle of the camera, and one on the left side of the camera near the top.

When the camera is recording, three red lights will appear—one recording light on the front of the camera and one in the back; on the side of the camera, a red light indicates which card is recording.

And of course, pressing the record button again stops the recording.

S & Q (SLOW-AND-QUICK MOTION)

The slow-motion capability of the FS7 is one of its best features, but it is not without its limitations.

The camera allows for slow-motion recording up to 60 frames per second (fps) in 4K and UHD and up to 180 fps in high definition or 1920 x 1080. The slow-motion footage is automatically processed in-camera, so it is rendered internally and can be played back immediately. A slow-motion clip is indistinguishable from any other clip on the card.

You can enable S & Q recording in two ways.

The first and easiest is by pressing the S & Q button on the upper left side of the camera.

The second way would be to navigate to:

RECORD > S & Q > ON
                    OFF

And turn it on.

Once you turn S & Q on, it will remain on until you turn it off. This is true even after you restart the camera, that is, if S & Q is on when you turn the camera off, when you turn the camera back on, S & Q will still be on. S & Q also works in both Custom and CINE-EI modes.

If you want to record slow motion at up to 180 frames per second aka high-frame-rate (HFR) recording, then you need to go to the Record menu and enable the necessary settings.

Given the high shutter speeds that come with shooting in slow motion, special consideration should be given to lighting and avoiding flicker. For these reasons, shooting under bright sunlight is often a good way to address these possible issues.

Note: You can only shoot HFR in high definition, that is, 1920 x 1080. The camera does not support HFR in UHD or 4K. UHD and 4K will only record slow motion up to 60 frames per second.

```
RECORDING > SETTING > ON

    > HIGH FRAME RATE MODE > OFF
                             FULL SCAN
                             CENTER SCAN

    > FRAME RATE > 72 - 180P
```

HIGH FRAME RATE MODE (Settings)

OFF	The maximum frames per second is 60. Full image scanned, no loss of quality.
FULL SCAN	Scans the entire sensor but artifacts may occur.
CENTER	Scans the center of the sensor and reduces the field of view by ½ but eliminates artifacts.

Under Frame Rate, if you choose the OFF Mode, you will see various frame rates between 1 and 60. If you choose either Full or Center scan, the frame rates will be various increments between 72 and 180.

When you are shooting in S & Q, your shutter speed should still be double your frame rate but this usually means you will need additional lighting to compensate for the faster shutter speeds.

KEY FUNCTIONS DISABLED DURING S & Q

AUTO FOCUS	AUDIO RECORDING
PICTURE CACHE	RELAY RECORDING

DISPLAY INDICATORS DISABLED DURING HFR

DEPTH OF FIELD	FOCUS ASSIST	FOCUS POSITION
IRIS POSITION	ZOOM POSITION	

The aim is to tell the truth without being boring. That's why lies are so successful because the truth can be pretty boring—because that's the truth.

MILOS FORMAN

SCAN MODE

USER > BASE SETTING > SCAN MODE >   NORMAL
                                    2K FULL
                                    2K CENTER

The sensor of the FS7 has 4,096 pixels across and 2,160 pixels down. This is an aspect ratio of 17 by 9 and is true 4k.

When shooting in ultra-high definition, the effective pixels are 3,840 by 2,160.

When you are shooting in normal mode, the sensor scans to its full range (Super 35 mm equivalent).

If you shoot in 2K Full, then the camera scans the full image, but converts it in camera and outputs 2K (2048 x 1080).

If you shoot in 2K Center, then the camera only scans and outputs the center section of the sensor (2048 x 1080 pixels).

When you shoot in 2K Center, it digitally crops the image and zooms in. In photographic terms, it visually cuts the field of view in half.

For example, if you are shooting with a 50 mm lens and are in normal mode, then the field of view is about 26 degrees. If you keep the 50 mm lens on and switch to 2K center, then you will have a field of view of about 13 degrees, which is equivalent to what you would get with a 100 mm lens.

In tests, there is no discernible drop in resolution or quality in 2K Center mode, and it cuts in seamlessly with other footage shot in Normal scan mode.

2K Center mode is a very helpful function if you need to zoom in on a subject but don't want to switch lenses or don't have a telephoto lens readily available. And it works out well because there is no loss of image quality when editing clips together.

SCENE FILES (Custom Mode only)

One feature the FS7 has is the ability to save and load scene files that allow you to change most of the Camera and Paint settings at once. This can be a major time-saver, especially if you have created a customized look in Custom mode and wish to save these settings for future use. However, using a scene file is not without its perils.

The scene file feature will save 79 out of 85 settings in the Paint menu and 24 settings out of 37 in the Camera menu. (Refer to pages 120-129 of the Sony FS7 manual for a complete listing of all the settings saved.)

This means that when a scene file is loaded into the camera, every one of those 103 settings (i.e., 79 + 24 = 103) will be imported into your camera. Unfortunately, when that happens, there are a couple of settings you probably don't want imported or changed.

> FILE > SCENE FILE > SCENE WHITE DATA > OFF

First, you probably do not want to import white balance values that were recorded under different lighting conditions. If that happens, it will overwrite any custom white balance values you have already set. Fortunately, you can prevent the scene file from importing white balance data by going to Scene White Data and turning it off.

Second, you probably don't want your shutter speed or angle for your current scene overwritten. Unfortunately, there is no way to prevent that data from getting imported, so once you load a scene file, you need to double-check your shutter speed or angle and make sure it didn't change.

When you are starting with the camera, creating your own scene files is not recommended until you are familiar with the camera and have all the tools you need.

If you do an Internet search using "FS7 scene files," two or three reputable sites will show up that offer free scene files for download.

Once you download the files, you transfer them to the utility SD card, then put the card into the camera, and load up the individual scene files.

SHUTTER

By the time you reach this setting, you should have already set your frame rate and should know what your shutter speed will be. Under this setting, you can choose the shutter mode and it can be set to speed, angle, ECS, or off.

In the table below, the dark blocks are for reference only and those shutter speeds and angles are not available. The white blocks mean that is an available setting on the camera.

SHUTTER SPEEDS (SS) & ANGLES (A)

24 FRAMES PER SECOND
SS	24	*28.8*	40	48	50	*57.6*	60	*72*	96	100	120	144	192
A	*360*	300	216	180	172.8	150	144	120	90	*86.4*	72	60	45

30 FRAMES PER SECOND
SS	30	*36*	50	60	*62.5*	*72*	75	90	100	*120*	125	180	*240*
A	*360*	300	216	180	172.8	150	144	120	*108*	90	*86.4*	60	45

60 FRAMES PER SECOND
SS	60	*72*	90	100	120	125	144	150	180	240	250	480	---
A	*360*	300	*240*	216	180	*172.8*	150	144	120	90	*86.4*	45	---

The shutter speed and angles available will depend on how many frames per second you are recording. Most of the shutter speeds and angles do not match each another. In referencing the table, the shutter speed matches the shutter angle 10 out of 38 times. In other words, only 26% of the shutter speeds and angles match.

If you set the shutter mode to ECS (Extended Clear Scan), you can set the frequency from 60 to 8000. This mode is helpful when you are shooting images of electronic displays and are trying to match the frequency of the device to get rid of horizontal scrolling bars.

One advantage of setting the shutter angle to 180 degrees is it will always double the shutter speed over the frame rate no matter which of the three frame rates you choose.

Also, you can adjust the shutter speed or angle quickly by pressing the Shutter button on the left side of the camera, but normally, shutter speed is a ***set-it-and-forget-it*** adjustment.

SLOT SELECTION

On the left side of the camera, there is a SLOT SELECTION button. If you press it, it changes the media card destined for recording. It is important to check this, so you can be sure you are recording to the correct card.

RECORDING > SIMUL

You also have the option under Recording Simultaneously to record to both cards at once. This is a useful feature if you are recording something of importance. Unfortunately, there are many instances where this function is disabled. Primarily, it only seems to work if you are shooting in high-definition (1920 x 1080) and at lower frame rates. Also, the function is disabled if S & Q or Picture Cache modes are enabled.

If you are recording with two cards (which is highly recommended), then the camera will automatically change slots just before a card fills up completely. Furthermore, do not remove a card if you see the red light on, that is, while it is actively reading or writing.

SWITCHING MODES

If you switch from CINE-EI to Custom mode, your gamma setting will change in Custom mode to what it was while you were in CINE-EI. For instance, if you were shooting in CINE-EI and using S-Log3, then when you switch back to Custom mode, the gamma will change to S-Log3.

You have two choices.

1. You can go into the menu and change your gamma mode back to what it was.

2. You can load up an All File with your preferred settings in it.

SYSTEM RESET

This setting returns your camera to all of its original default settings. If you are using a FS7 for the first time or are using it after someone else, then a system reset is recommended. This way you know you are starting from the default settings.

In addition, if you should find that the camera is acting strange or controls are not behaving as they should, the System Reset function is frequently helpful in resolving whatever the glitch might be.

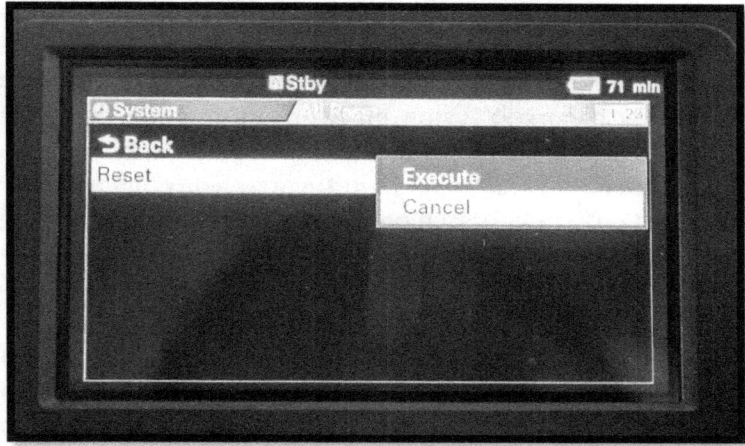

VF (VIEWFINDER)

VIEWFINDER MENU OPTIONS

VF SETTING	VF brightness and color.
PEAKING	Focus peaking settings.
ZEBRA	Set zebras for IRE measurement.
MARKER	Controls layout, frame markers, etc.
DISPLAY ON/OFF	Set what items appear on the display.
VIDEO SIGNAL MONITOR	Type of scope to display on viewfinder.

In Custom mode, it is recommended that you display the waveform monitor on the viewfinder. Though the monitor is small, it can be a handy visual reference at times. Usually, it can provide a quick confirmation that your white levels remain where they need to be.

```
VF (VIEWFINDER) > VIDEO SIGNAL MONITOR > SETTING > OFF
                                                  WAVEFORM
                                                  VECTOR
                                                  HISTOGRAM
```

Under Video Signal Monitor, you can enable which scope appears on your viewfinder. You have three choices: Waveform, Vector, and Histogram. Of the three choices, the waveform monitor is probably the most useful.

In Custom mode, all Monitoring LUTS are disabled, so you don't have to deal with any MLUT settings in Custom mode. What you are seeing on the viewfinder is what you are recording. The only exception to this is shooting with a S-Log gamma in Custom mode. In this situation, the image you are seeing on the viewfinder might not be an accurate representation of what you are actually recording, but unfortunately, you don't have any built-in Monitoring LUTS in Custom mode. Some external monitors have Monitoring LUTS, but you would have to configure those according to the monitor's manufacturer.

In CINE-EI mode, you will also want to display the waveform monitor on the viewfinder, and you would set it the same way you would in Custom mode. The main concern here is that you need to pay attention to which signal is being monitored by the waveform monitor.

VIDEO

In this section, you can control the video output being sent to the external HDMI output and the two SDI outputs on the right side of the camera. Different monitors have different configurations, so if you want to attach an external monitor, you will need to refer to the monitor's manufacturer for more information about how to configure the settings.

There are a few important things to note.

```
VIDEO > OUTPUT FORMAT >
                (SDI)              (HDMI)
                > - - -            > 4096 x 2160P
                > - - -            > 3840 x 2160P
   (Level A)    > 1920 x 1080P     > 1920 x 1080P
   (Level B)    > 1920 x 1080P     > - - -
                > 1920 x 1080i     > 1920 x 1080i
```

If the video output is set to 4K (4096 x 2160) or UHD (3840 x 2160), you cannot use the scopes to monitor S-Log and will be limited to monitoring the MLUT only.

```
VIDEO > MONITOR LUT > SDI1 & INTERNAL > MLUT OFF
                    > SDI2  > MLUT OFF
                    > HDMI  > MLUT OFF
                    > VIEWFINDER  > MLUT ON
```

You must be careful that you do not turn the MLUT setting ON for the SDI1 and INTERNAL option. If you do, then the MLUT-look that you are monitoring on the viewfinder with be recorded internally to your memory cards and overwrite the S-Log footage. This will permanently alter your footage in a way you might not want. (If you want to bake-in looks to your footage, switching to Custom mode and making use of the Paint menu would be better than baking in a look using a MLUT in CINE-EI mode.)

Fortunately, the scope identifies what signal is being monitored, and it appears just above the scope.

VIDEO > OUTPUT > SDI > ON/OFF

This setting doesn't affect the scope display on the viewfinder or the ability for the scopes to monitor the recording signal; it affects whether there is a signal going to the SDI external output.

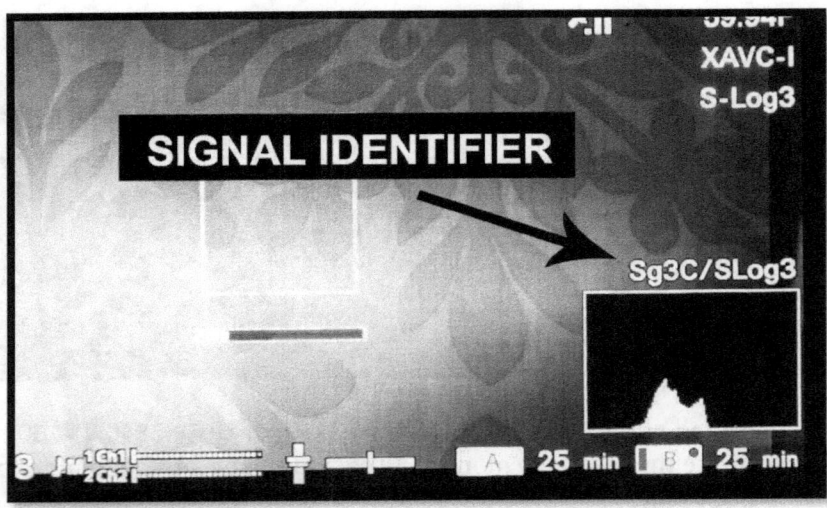

There is a signal identifier that appears above the scope to let you know which signal is being monitored. In the picture above, it indicates that the S-Log3 video signal is being monitored. If the MLUT were being monitored instead, it would indicate LUT 709. This is a good feature because if you inadvertently change an output setting, the signal identifier will alert you that the wrong signal is being monitored, and you can adjust the settings to resolve it.

> *If you don't get physically ill seeing your first rough cut, something's wrong.*
>
> **MARTIN SCORSESE**

VIDEO SIGNAL MONITOR

```
VF (VIEWFINDER) > VIDEO SIGNAL MONITOR > SETTING > OFF
                                                 WAVEFORM
                                                 VECTOR
                                                 HISTOGRAM

                              > SOURCE > SDI1 & INTERNAL REC
                                         SDI2
```

With this setting, you can enable which scope displays on the viewfinder. You have three choices: Waveform, Vector, and Histogram. The waveform monitor is probably the only scope you will need. Since the scope is so small, it only gives an approximation of the IRE levels.

In CINE-EI, if the Source is set to **SDI1 & Internal Recording** *and* the **Video > Monitor LUT > SDI1 & Internal Rec >** is On, then the scopes will monitor the MLUT. If the Source is set to SDI2, then the scopes will monitor the underlying S-log whether **Video > Monitor LUT > SDI1 & Internal Rec** is off or on.

In Custom mode, the Source is locked to the Internal Recording.

WHITE CLIP

This sets the level at which highlights or whites will show as clipping.

By default, it is set at 108%.

Under this menu option, you can turn the setting on or off, then set the clip level anywhere from 90 to 109%.

It is recommended that you turn the function on and set the clip level to its maximum of 109%. This will allow the maximum range possible.

The white clip adjustment is not available when you are shooting in S-Log2 or 3.

XQD MEMORY CARDS

The FS7 uses proprietary XQD memory cards. The cards are bigger than SD cards and a little smaller than Compact Flash cards. The cards are reported to have a number of advantages over the other types of cards and are optimized for the FS7.

There are two main series of the card available, the G and M series. The G and M series have much in common; however, the G series covers every format the camera records internally and has faster transfer speeds when writing, so it is recommended that you get a G-series card.

The G series is available in 32, 64, 128, and 256 GBs.

If you are shooting in 4K, not only is recommended that you get a G series card, but it is also suggested that you get a large-capacity card, for instance, 128 or 256 GB. According to Sony, the 256 GB model allows up to approximately 45 minutes when shooting high-quality 4K.

It should be noted that since these cards are proprietary, you will need to get a special reader for them.

It can be difficult to eject XQD cards from the FS7. One suggestion is to push on the card before you eject it, then quickly slide your finger off while still pushing down. This will cause the card to pop out farther than it normally does and should make it easier to pull out. Also, if you don't like ejecting the cards, you can always use a USB cable to access them.

ZEBRA

The best way to get a proper exposure is by using the zebra function on the camera.

The Zebra function is divided into five sections.

```
USER > ZEBRA > SETTING > ON
                        OFF

              > ZEBRA SELECT > 1
                             > 2
                             > BOTH

              > ZEBRA 1 LEVEL > 0-107%

              > ZEBRA 1 APERTURE LEVEL > 1 – 20%

              > ZEBRA 2 LEVEL > 0 -107 %
```

First, you can set Zebra 1, 2, or both. It is recommended that you enable Zebra 1 and not enable both.

Zebra 1 can be used when you are trying to be precise in your exposure assessment. Normally, you set your Zebra 1 to either the middle gray value or the 90% white reflectance level. For accuracy, you can set the level of accuracy to 3%. The aperture setting controls the range at which zebras will begin to appear. For instance, if you set Zebra 1 to 40% with an aperture of 3%, the Zebra will appear in the range from 38.5 to 41.5%. If you set Zebra 1 to 75% and set the aperture to 10%, then zebras will appear from 70 to 80%. Setting the aperture to 3% is usually a safe bet.

Zebra 2 can be viewed more as an overexposure system but can also be used for exposure as well. Some people prefer using it to Zebra 1 because it triggers at the exact threshold you set, and it stays on so long as you remain at or above that threshold.

> *Film is truth 24 times a second, and every cut is a lie.*
> **JEAN-LUC GODARD**

BASIC COLOR CORRECTION
(USING DA VINCI RESOLVE 14)

STEP 1 / APPLY A LUT

If you shot with S-Log, then the first step in the process is to apply a LUT. Applying the LUT first makes sense because it is possible that you will love the way the LUT transforms the footage. And if you really like what the LUT does, then you might not need to make any further changes! But that's probably not going to happen, and you will want to make other adjustments.

The main thing is being sure to apply the right LUT. If you shot in S-Log2, then you want to be certain to choose a S-Log2 LUT that converts the imagery to REC-709; if you shot in S-Log3, then make sure you choose a S-Log3 LUT to convert it to REC-709.

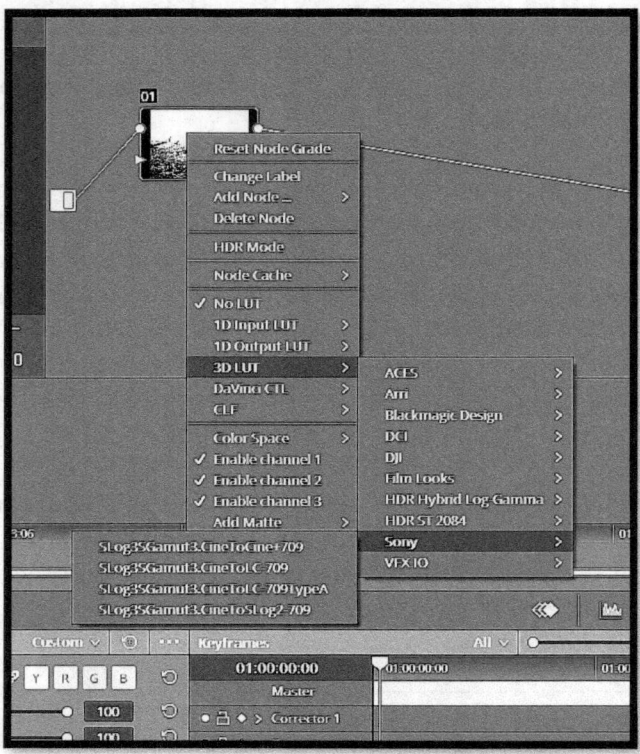

The first serial node you create should be for the LUT. After you apply the LUT to that node, then you will decide whether to put additional serial nodes either before or after it. It is advisable to only apply one type of adjustment per node.

The rule is: If you like the LUT, then you will want to work within the parameters of the LUT; therefore, any additional corrections should be applied on serial nodes that you will place before the LUT.

If you want to change the LUT's overall appearance and work beyond the constraints it applies, then you will want to place your serial nodes after the LUT.

And if you hate the LUT, you can always delete it and try another.

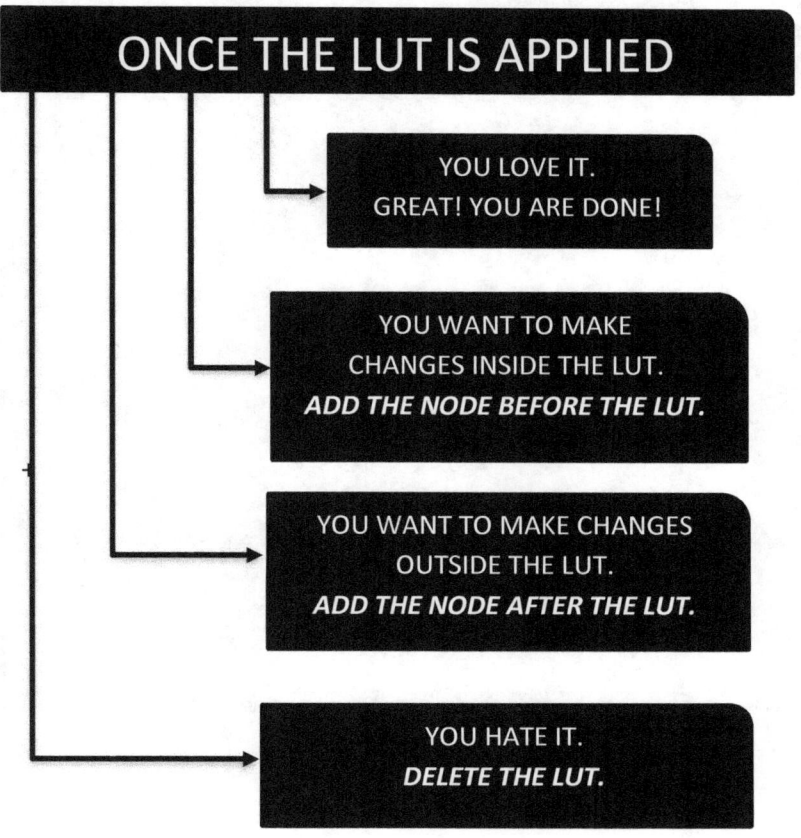

STEP 2 / LIFT

The second step in the process is to adjust the black level. You can think of the black level as the floor or foundation of your project. If the base or floor is off, then it can throw the look of the entire image off.

The question here is how dark do you want the blacks? Do you want deep blacks with no detail or do you want softer blacks with more shadow detail?

If you shot in S-Log, you should use the Log Wheels to make your adjustments. This will result in more precise control over your shadows, mid-tones, and highlights; if you shot with any other non-S-Log curves, you can use the Primary wheels.

To begin this adjustment, you should pull up the waveform monitor and position it so it isn't in the way of anything important. Whether you are adjusting lift or gamma, you should not pull blacks below 0 IRE or push whites above 100 IRE.

On Da Vinci Resolve, the waveform scale is in code values, not IRE. You need to know that on this scale: **100 IRE = 940** and **0 IRE = 64**. You can enable reference lines at these values by clicking the slider icon in the upper-right corner of the monitor's window.

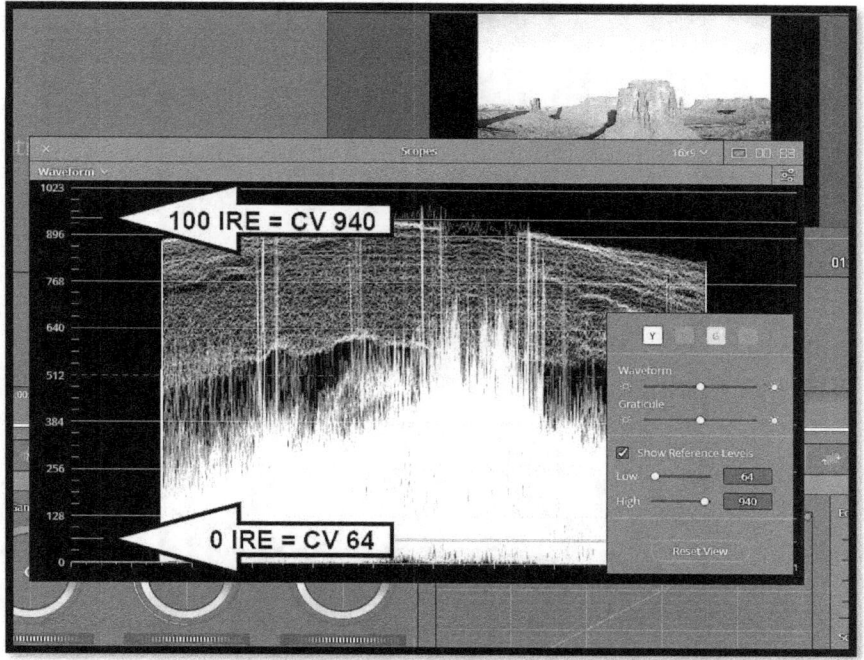

It should be noted 90 IRE is a code value of 855. Often this is a better spot for whites than 100 IRE.

CODE VALUES		
	CODE VALUES	IRE %
BLACK	64	0
MIDDLE GRAY	414	41.7
WHITE	940	100

Once the waveform monitor is setup, then apply another serial node before the node the LUT is on. Once the node is added, then adjust the lift (aka shadows) and keep an eye on the waveform and the image. Adjust the dial to bring the blacks up or down to your liking.

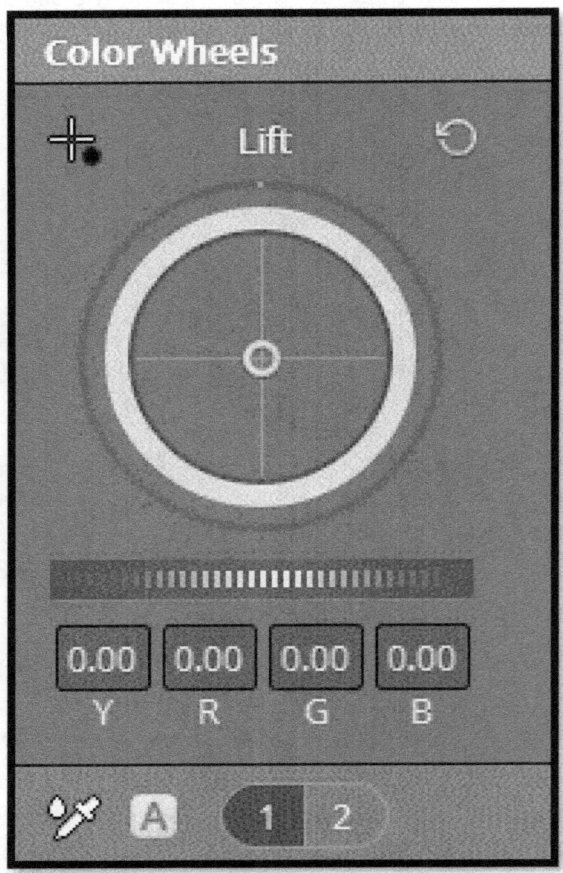

As a rule, you don't want to adjust the blacks below 64 on the waveform monitor. Beyond 0 IRE or 64 CV, black doesn't get any blacker.

STEP 3 / GAIN

Next assess the image for highlights and clipping. If the highlights need to be brought down, then adjust the gain accordingly. You can keep the gain adjustment on the same node as the lift adjustment.

The rule here is to consider bringing your brightest whites down to 855 or 90 IRE on the waveform monitor. Taming highlights is often associated with a cinematic look. But again, this is a subjective assessment, and the look you want is a creative choice, not a scientific one.

If you wish to adjust the overall exposure level of the image and bring lift, gamma, and gain up or down all at once or in unison, then you can use the offset dial for that.

STEP 4 / GAMMA

Now look at the mid-tones and assess the overall brightness of the image. Also note where skin tones are falling on the waveform monitor. If you want to increase or decrease the mid and skin tones you can do that here. You can also keep this adjustment on the same node as lift and gain.

RANGES FOR SKIN TONE		
(approximate)		
SKIN TYPES	DA VINCI CODE VALUES	IRE %
LIGHT	502 - 677	50-70%
MEDIUM	326 - 502	30-50%
DARK	239 - 502	20-50%

The range of luminance for skin will vary depending on skin type, and the above table is offered as an approximation only. Men usually fall about 5-10% lower than women in terms of skin reflectance.

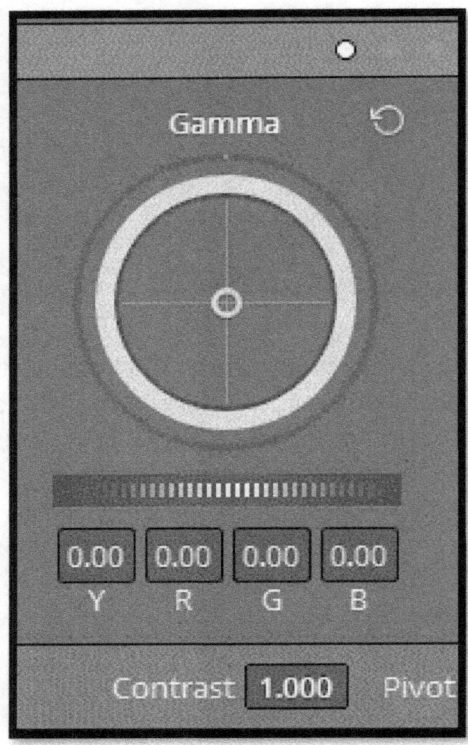

STEP 5 / OFFSET

With a LUT applied and the lift, gamma, and gain adjustments complete, the image should be coming together; however, if the image still looks under or overexposed, you can darken or brighten the entire image by adjusting the offset. And if that's the case, then you can make the offset adjustment now and keep it on the same node with the lift, gamma, and gain.

The offset function will treat lift, gamma, and gain as one and bring them all up or down at the same time.

At this point, you should have one node with a LUT applied and one node before it with your lift, gamma, gain, and offset adjustment.

For convenience only, the remaining three adjustments will be added together on a single node. But keep in mind it is usually recommended that you keep your adjustments on separate nodes. This is done so you can better manage the corrections you make and isolate them if needed.

STEP 6 / COLOR BALANCE

Once the image looks properly lit, you can assess the white balance. To begin, pull up the RGB parade to see if the core structures of R, G, and B channels are in alignment.

If not, follow this procedure.

1. Go to Workspace > Video Scopes and display the RGB Parade.

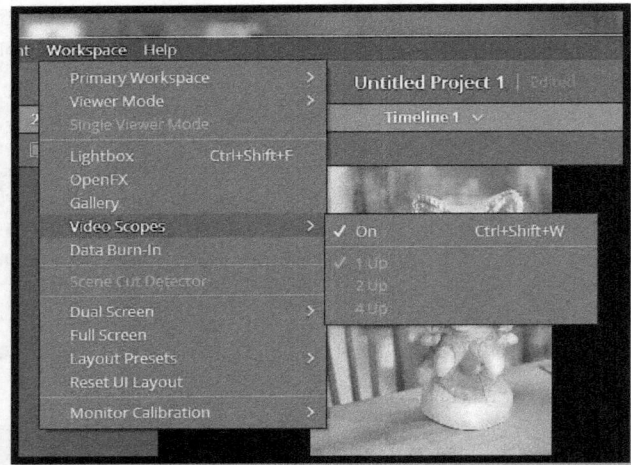

2. Change from Primaries Wheels to Primaries Bars.

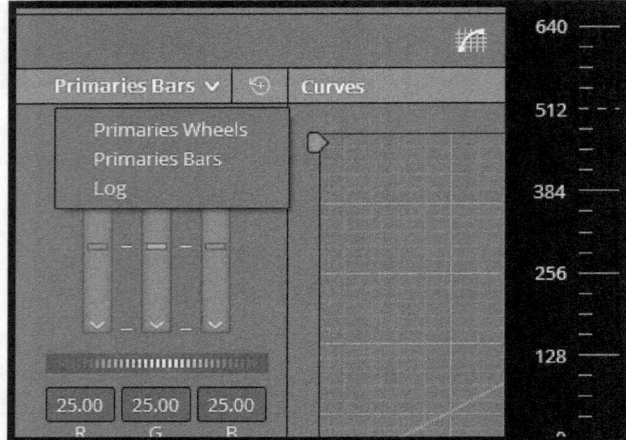

3. Add a serial node by right clicking an existing node.

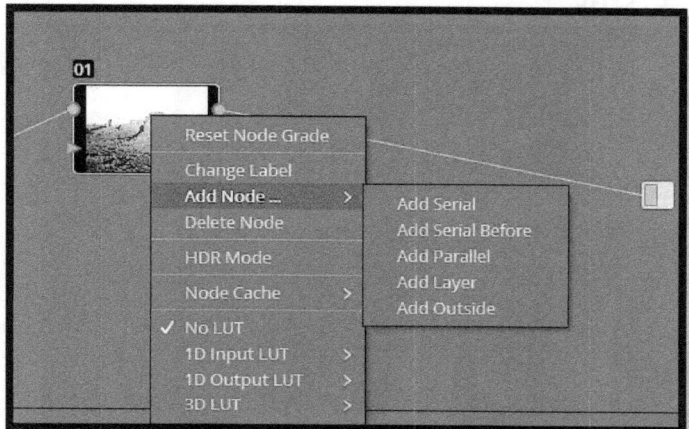

Nodes can also be added from the Main Menu under Nodes.

4. Under Gain, adjust each of the color channels (R,G,B) individually. Try to get the channels as evenly aligned as possible.

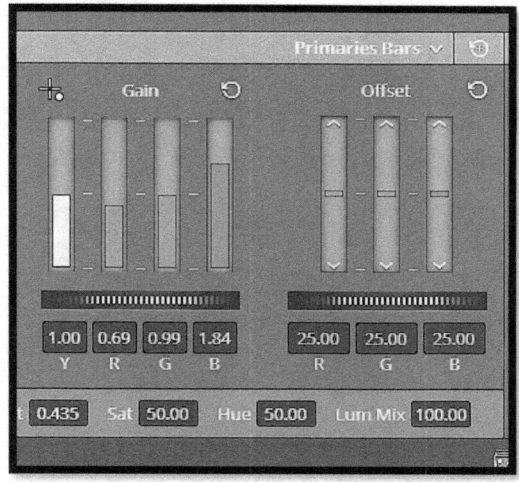

Start with the red channel, then green, and finally blue. You'll notice as you adjust one color, it can affect the other colors. The channels might not align perfectly and that's to be expected, especially in mixed lighting conditions. Some colors will spike and some colors won't even have spikes, so it will be impossible to match 100%.

BEFORE

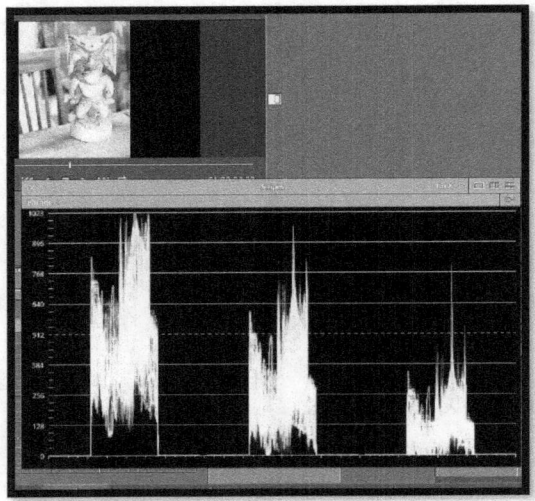

AFTER

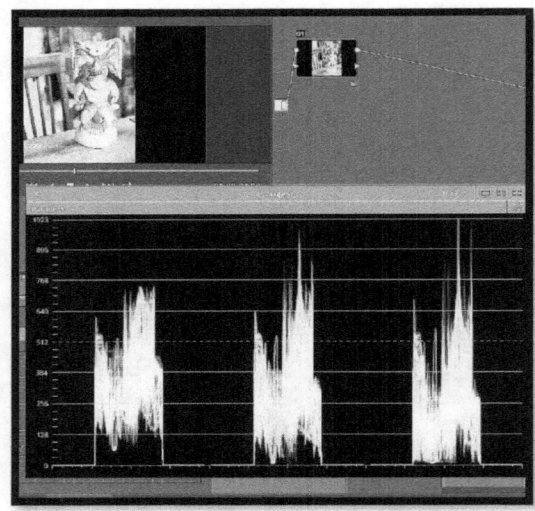

Notice in the above picture how the blue channel (the one on the far right) has spikes and the red channel (the one on the far left) does not. The main thing is to align the core structure of each channel's shape. And be sure to look at the image itself and not just the RGB Parade.

Ultimately, you must trust your eyes to determine the exact look you want.

STEP 7 / CONTRAST

With the lift, gamma, gain, and white balance done, you might want to consider adding contrast to the overall image. Contrast will affect the entire image and creates a stronger and more striking image. However, this adjustment, like the others, is subjective, so it is worth experimenting with to see if you like its effect. You adjust the contrast by placing the mouse cursor over the number by the Contrast control, then a double-sided arrow will appear. Simply click and drag to decrease or increase the contrast.

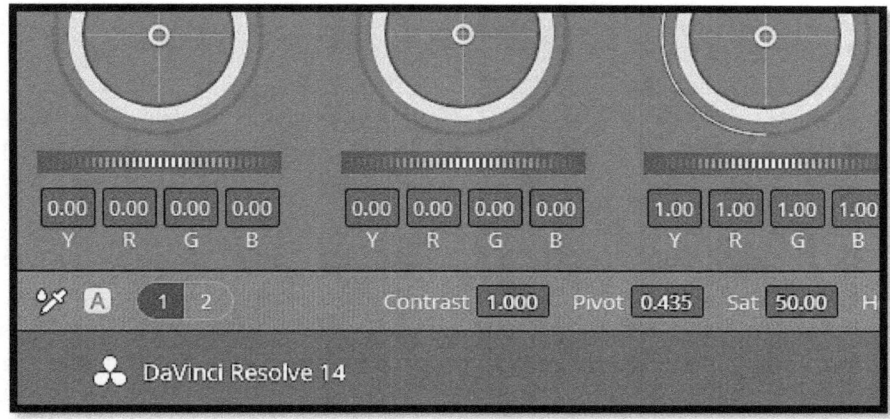

STEP 8 / SATURATION

The final step is to play with the saturation dial and see if you want to decrease or increase the intensity of colors. The saturation adjustment works the same way as contrast does.

There is much more to grading than what is presented here.

The Da Vinci Resolve manual is available for free online, and you are encouraged to download it to learn more about the program. But note that as of May 2018, it is 1,346 pages long. This section only touches on a very small part of what the program is capable of doing.

KEY CONCEPTS

- Anything shot in CINE-EI must be graded in post.

- Anything shot in CINE-EI mode can be graded to look like anything shot in Custom mode, but not everything shot in Custom-mode can be graded to look like CINE-EI footage.

- In CINE-EI, S-Log2 and S-Log3, no matter the color space, can be graded to look virtually identical.

- Getting skin tones right is at least half the color correction process.

- Consistent exposure levels across scenes will make matching shots much easier in post.

- The more you can get right in camera, the better it is for the final footage.

- With S-Log, applying a LUT is usually the most efficient and productive way to get started.

- Color correction is not tensor algebra but does take time and practice to master.

- Many decisions made in the grading process can be considered artistic and completely subjective.

> *But the cinematographer as a technician must also have a deep understanding of the tools available for use today. He or she must research and test with every camera available to widen his or her technical knowledge of the capabilities of the systems available and, more important, to understand their shortcomings.*
>
> **DAVID STUMP**

ONLINE RESOURCES

ABEL CINE [scene files]
https://www.abelcine.com/articles/blog-and-knowledge/tools-charts-and-downloads/new-sony-fs7-scene-files-from-abelcine

B & H PHOTO [Lastolite EzyBalance Gray Card]
https://www.bhphotovideo.com/

CUE 93 STUDIOS [noise reduction profiles for Neat Video]
http://www.cue93.com/a-look-at-noise-reduction-and-sony-fs7-firmware-v3-1/

DIY DIGITAL CINEMA [book updates]
http://diydigitalcinema.com/

DVX USER [active FS7 user forum]
http://www.dvxuser.com/V6/

SONY [manual, firmware, brochures, product information]
https://pro.sony/ue_US/products/handheld-camcorders/pxw-fs7#ProductResourcesBlock-pxw-fs7

XDCAM-USER [scene files]
http://www.xdcam-user.com/2015/02/scene-files-picture-profiles-for-the-pxw-fs7/

YOU TUBE [Sony training videos for FS7]
https://youtu.be/h_GBCykaSmE

DEFINITIONS

4K — *A display resolution of at least 4,000 pixels along one axis. The FS7 is a true 4K camera capable of recording an image that is 4,096 by 2,160 pixels.*

Adaptive Matrix — *The adaptive matrix minimizes excessive blue in an image. Excessive blue can occur under LED lights.*

APR — *Automatic pixel restoration.*

Chroma — *Chroma is short for chrominance, which is the color information in a signal. It refers to how colorful something is in relation to something bright or white.*

CINE-EI — *An exposure index or system that precisely calibrates and controls a monitoring LUT, which can assist in determining exposure adjustments.*

Codec — *Short for **co**mpression-**dec**ompression. It is a system for compressing and decompressing digital data for practical and streaming purposes.*

Gain — *In Da Vinci Resolve, gain refers to highlights. In electronics, it refers to signal amplification.*

Gamma — *In Da Vinci Resolve, gamma refers to midtones or the middle area between highlights and shadows. In digital video, it is a numeric representation of luminance.*

Histogram	*A bar graph representation of luminance. Blacks are on the left and whites on the right.*
Hue	*Another word for color.*
Hypergamma	*A method of encoding luminance to maximize the visible dynamic range. It redistributes visual information in a way that takes advantage of human perception, which is more sensitive to shadows than highlights.*
Lift	*In Da Vinci Resolve, this is another word for shadows.*
Look Profile	*Another term for a 3D-Lut.*
Luminance	*The brightness or "the black, white, and range of gray" information in a signal.*
LUT	*Look-up table. A mathematical transformation of digital information from one space to another.*
Multi-Matrix	*This a submenu within the Paint menu of the FS7. It allows the user the ability to precisely control how colors are recorded and subsequently reproduced.*
Neat Video	*A noise-reduction plug-in. It is compatible with most major video editing programs.*
Offset	*In Da Vinci Resolve, this is a de-facto exposure control. It allows the user the ability to lighten or darken an entire image at once.*
Phase	*A setting in Matrix menu that shifts colors along their axes.*

Post	*Short for post-production. This usually refers to adding effects, color correction, and editing.*
Primary correction	*Additions and adjustments that affect an entire image. In video, this would include applying a LUT, exposure, white balance, and contrast adjustments.*
Pro Res	*A high-quality codec developed by Apple. ProRes is considered visually lossless, which means that even though the image is compressed and data is lost, the effect on the video is negligible.*
Rec. 709	*A display and color-space standard first introduced in 1990 for high-definition television (HDTV).*
Rec. 2020	*A display and color space standard for ultra-high-definition television (UHDTV).*
RGB Parade	*Three waveform displays next to each other as if in a parade. One waveform is for the red channel, one for green, and one for blue.*
S Gamut	*The original Sony color space. It exceeds the Rec. 2020 color space.*
S Gamut3	*The revised Sony color space. It is very close to the color space of the FS7's sensor and also exceeds the Rec. 2020 color space.*
Saturation	*The relative intensity of chroma.*
Secondary correction	*Additions and adjustments that affect only part of an image. In video, this would include applying masks, isolating and enhancing skin tones, and making other selective corrections.*

S-Gamut3.cine	The smallest of the available color spaces. It is wider than the DCI-P3 space, which is the cinematic standard.
S-Log	First-generation gamma curve developed by Sony. Not available on the FS7.
S-Log2	Second-generation gamma curve. Fourteen stops of dynamic range. Clips at 107 IRE.
S-Log3	Third-generation gamma curve. Mimics the Cineon standard for motion-picture film scanning and Arri LogC. Clips at 93 IRE.
UHD	Ultra-high-definition. It has a resolution of 3840 x 2160 pixels. Sometimes this is mistakenly referred to as 4K but falls short of true 4K.
Vectorscope	Measures chroma and hue in a video signal.
Waveform	Graphical representation of luminance from a scene or image. Useful as an exposure tool.
XAVC	A proprietary video-recording format developed by Sony in 2012. It can record 4K up to 60 frames per second, supports color depths up to 12 bits, and chroma subsampling up to 4:4:4. It uses the Material Exchange Format (.mxf) as its wrapper or digital container. It can also record up to 8 channels of audio.
XDCA	Extension unit that attaches to the back of the FS7. Enables 12-bit 4K/2K RAW data output, time code, 1080p ProRes internal recording, and the attachment of V-mount batteries.

The following checklists have been condensed, so that they can be discreetly carried and used in the field.

Please feel free to photocopy them for your use.

If the checklists will be seeing a lot of use, it would also be advisable to laminate them.

QUICK START CHECKLIST

Install lens and leave the cap on.
Install XQD card(s).
Install fully charged battery.
Power on, and check Hold switch is off.
If APR activates, stop down the iris and execute it.
Jump to: System > Version---Is firmware current? 4.20
System > All Reset---Reset camera.
System > HOLD Switch Setting > w/ Rec---Turn off.
System > Language: English---Change if needed.
System > Clock Set---Change as needed.
Turn off all Auto Settings.
Turn off Auto Focus.
Return to: User > Country---Set PAL or NTSC.
User > Base Setting > Shooting Mode---Set to Custom.
User > Rec Format---Select format.
User > Codec---Select.
User > Assignable Buttons---Assign buttons 2 and 3.
User > Format Media---Format.
External: Confirm Slot Selection.
User > Gamma---Select.* Check chart, note white 90% level.
User > Aperture---Turn on. Set to +20.
User > Peaking---(Optional setting.)
User > ISO / Gain---Set ISO or dB. Set values L, M, & H.
User > Zebra---Set level based on gamma.
Camera > Shutter---Set mode. Set speed/angle.
Camera > Color Bars > Setting ---Turn on, then choose:
Type > SMPTE---Calibrate contrast and brightness.
Camera > Auto Black Balance---Execute.
Paint > Black Level---Set to -3.
Paint > White Clip---Set to 109.
External: AUTO/MAN (Audio)---Set to Manual (CH 1 & 2).
Audio > Audio Inputs---Set inputs, wind filter on, side level.
Audio > Audio Output > Monitor Volume---Adjust.
VF > Video Signal Monitor > Setting---Choose waveform.
Check display on VF to confirm all settings.
Remove lens cap.
Make sure lens is clean, free from dust and smudges.
Install lens hood or matte box.
Confirm the ISO switch is set to L.
Adjust exposure.
Set WB switch to A or B. Perform custom WB.
Adjust focus.
Check audio levels on Display or Status page.
Put settings on hold to prevent changes. (optional)
Ready to record

CUSTOM MODE CHECKLIST

Install lens and leave the cap on.
Install XQD card(s) and fully charged battery.
Power on and make sure the Hold switch is off.
If APR activates, stop down the iris and execute it.
Turn off all Auto Settings.
Turn off Auto Focus.
Set: Base Setting > Shooting Mode > Custom.
Set: Base Setting > Scan Mode. Change if needed.
Set: Rec Format, take mental note.
Set: Codec. Format Media. Confirm Slot Selection.
If loading a scene file:
Jump to: File > Scene File > Scene White Data > Turn off.
Then load the scene file and return to User menu.
Gamma* : Be sure to remember what this setting is.
Aperture > On > Set to 10-20.
Peaking...Optional.
ISO / Gain / EI * : Set values for L, M, & H.
Set Zebra 1 based on gamma.
Shutter speed / angle. *
Noise suppression.* Set to high in low light.
Auto Black Balance. Execute.
Black* : -3 (optional) / White Clip* : 109
Paint > Matrix > Setting* : On
Paint > Matrix > Adaptive Matrix* : On
Paint > Matrix Preset* : (optional)
Paint > Preset Select* : (optional)
Paint > User Matrix* : (optional)
Paint > Matrix User > Level* : (optional)
Paint > Multi-Matrix > Setting* : (optional)
Paint > Multi-Matrix* : YL and R set to -20 to -30. (optional)
Audio Switch: Set to Manual.
Audio: Set inputs, wind filters on, side level.
Audio Output> Adjust monitor volume.
VF > Video Signal Monitor > Setting > Waveform.
Use display to confirm all settings.
Remove lens cap.
Make sure lens is clean, free from dust and smudges.
Install lens hood.
Confirm the ISO switch is set to L.
Adjust exposure.
Set WB switch to A or B. Perform custom WB.
Adjust focus.
Check audio levels. Display or status.
Put settings on hold to prevent changes. (optional)
Ready to record.   [* *Scene file loads setting.]*

CINE-EI CHECKLIST

Install lens and leave the cap on.
Install XQD card(s).
Install fully charged battery.
Power on.
Undo the hold setting if on.
If APR activates, stop down and execute.
Turn off all Auto Settings.
Turn off Auto Focus.
Set: Base Setting > Shooting Mode > CINE-EI.
Set: Base Setting > Color Space. Choose one.
Set: Base Setting > Scan Mode. Choose Center to zoom.
Set: Rec Format, take mental note.
Set: Codec
Format Media
Confirm Slot Selection.
ISO / Gain / EI : Set values for L, M, & H.
Set zebra based on gamma
Set: Shutter speed or angle.
Noise suppression. Set to high if low light.
Execute: Auto Black Balance
Audio Switch: Set to Manual
Audio: Internal, Wind, Side Level
Audio Output> Monitor Volume, Headphone Out, Output
Video > Monitor LUT > SDI1 & Internal > Off. Confirm off.
Video > Viewfinder > MLUT > Turn on.
Set: Video Signal Monitor > Setting > Waveform.
Set: Video Signal Monitor > Source > SDI2.
Use display to confirm all settings.
Remove lens cap.
Make sure lens is clean, free from dust and smudges.
Install lens hood.
Confirm the ISO switch is set to L.
Adjust exposure
Set white balance closest to lighting.
Adjust focus.
Check audio levels. Display or status.
Put settings on hold to prevent changes. (optional)
Ready to record.

INDEX

4K	80
ADAPTIVE MATRIX	69
ALL FILE	16
APERTURE	17
APR	17
ASSIGNABLE BUTTON	18
AUDIO	19
AUDIO LEVELS	22
AUDIO OUTPUT	23
AUTO BLACK BALANCE	24
AUTO FOCUS	24
AUTO SETTINGS	25
BASE SHOOTING MODE	26
BATTERIES	27
BATTERY LIFE	27
BLACK LEVEL	28
CALIBRATION	29
CAMERA OUTPUTS	30
CHANGING CHECKLIST	15
CHANNEL 1 & 2 CONTROL	30
CHROMA	75
CINE-EI CHECKLIST	14-15
CINE-EI MODE	31
CLOCK SET	32
CODEC	32
COLOR SPACE	33
COUNTRY (NTSC/PAL)	34
CUSTOM MODE	35
CUSTOM WHITE BALANCE	36-38
CUSTOM-MODE CHECKLIST	12-13
DA VINCI RESOLVE	94-105
DISPLAY	39-40

EXPOSURE	41
FIRMWARE	48
FLICKER	49
FOCUS	49
FOCUS ASSIST	50
FORMAT MEDIA	50
GAIN	99
GAMMA	100
HDMI	30
HIGH / LOW KEY	55
HISTOGRAM	91
HOLD FUNCTION	55
HUE	72
HYPERGAMMA	52-54
INTERVAL (TIME-LAPSE)	56
ISO / GAIN / EI	57-62
ISO / GAIN / EI SWITCH	63
LANGUAGE	64
LENS / CAP / HOOD	64
LIFT	96
LOOK PROFILE	65
LUT	65
MAPPING LUTS	67
MATRIX	69
MONITOR LUT (MLUT)	70
MULTI-MATRIX	71
NOISE SUPPRESSION	73
OFFSET	101
PAINT	74
PEAKING	72
PHASE	75
POWER	78
PRESET WHITE BALANCE	79-80
PRIMARY CORRECTION	110
PRO RES	32

QUICK-START / INITIAL LIST	10
REC 2020	63
REC 709	52
REC FORMAT	80
RECORD	80
RGB PARADE	104
S & Q	81-82
SATURATION	65, 71, 75
SCAN MODE	83
SCENE FILES	84
SDI	30
SECONDARY CORRECTION	110
S-GAMUT	33
S-GAMUT3	33
S-GAMUT3.CINE	33
SHUTTER	85
S-LOG2	33, 45
S-LOG3	33, 45
SLOT SELECTION	86
SWITCHING MODES	87
SYSTEM RESET	87
UHD	80
USER MATRIX	69
VECTORSCOPE	91
VF (VIEWFINDER)	88
VIDEO	89-90
VIDEO SIGNAL MONITOR	91
WHITE CLIP	91
XAVC	32
XDCA	28, 31, 32
XQD MEMORY CARDS	92
ZEBRA FUNCTION	93

NOTES

www.ingramcontent.com/pod-product-compliance
Lightning Source LLC
Chambersburg PA
CBHW070147230526
45471CB00002B/553